Undergraduate Texts in Mathematics

T0351055

Undergraduate Texts in Mathematics

Apostol: Introduction to Analytic
Number Theory.
1976. xii, 338 pages. 24 illus.

Armstrong: Basic Topology.
1983. xii, 260 pages. 132 illus.

Bak/Newman: Complex Analysis.
1982. x, 224 pages. 69 illus.

Banchoff/Wermer: Linear Algebra
Through Geometry.
1983. x, 257 pages. 81 illus.

Childs: A Concrete Introduction to
Higher Algebra.
1979. xiv, 338 pages. 8 illus.

Chung: Elementary Probability Theory
with Stochastic Processes.
1975. xvi, 325 pages. 36 illus.

Croom: Basic Concepts of Algebraic
Topology.
1978. x, 177 pages. 46 illus.

Curtis: Linear Algebra:
An Introductory Approach.
1984. x, 337 pages. 37 illus.

Dixmier: General Topology.
1984. x, 140 pages. 13 illus.

Ebbinghaus/Flum/Thomas
Mathematical Logic.
1984. xii, 216 pages. 1 illus.

Fischer: Intermediate Real Analysis.
1983. xiv, 770 pages. 100 illus.

Fleming: Functions of Several Variables.
Second edition.
1977. xi, 411 pages. 96 illus.

Foulds: Optimization Techniques: An
Introduction.
1981. xii, 502 pages. 72 illus.

Foulds: Combinatorial Optimization for
Undergraduates.
1984. xii, 222 pages. 56 illus.

Franklin: Methods of Mathematical
Economics. Linear and Nonlinear
Programming. Fixed-Point Theorems.
1980. x, 297 pages. 38 illus.

Halmos: Finite-Dimensional Vector
Spaces. Second edition.
1974. viii, 200 pages.

Halmos: Naive Set Theory.
1974, vii, 104 pages.

Iooss/Joseph: Elementary Stability and
Bifurcation Theory.
1980. xv, 286 pages. 47 illus.

Jänich: Topology
1984. ix, 180 pages (approx.). 180 illus.

Kemeny/Snell: Finite Markov Chains.
1976. ix, 224 pages. 11 illus.

Lang: Undergraduate Analysis.
1983. xiii, 545 pages. 52 illus.

Lax/Burstein/Lax: Calculus with
Applications and Computing, Volume 1.
Corrected Second Printing.
1984. xi, 513 pages. 170 illus.

LeCuyer: College Mathematics with
A Programming Language.
1978. xii, 420 pages. 144 illus.

Macki/Strauss: Introduction to Optimal
Control Theory.
1981. xiii, 168 pages. 68 illus.

continued after Index

Winfried Scharlau
Hans Opolka

From Fermat to Minkowski
Lectures on the Theory of Numbers and
Its Historical Development

With 31 Illustrations

Springer-Verlag
New York Berlin Heidelberg Tokyo

Winfried Scharlau
Universität Münster
Mathematisches Institut
Fachbereich 15 Mathematik
4400 Münster
West Germany

Hans Opolka
Universität Münster
Mathematisches Institut
Fachbereich 15 Mathematik
4400 Münster
West Germany

Translators

Walter K. Bühler
Springer-Verlag
New York, NY 10010
U.S.A.
:

Gary Cornell
Department of Mathematics
University of Connecticut
Storrs, CT 06268
U.S.A.

Editorial Board

F. W. Gehring
Department of Mathematics
University of Michigan
Ann Arbor, MI 48109
U.S.A.

P. R. Halmos
Department of Mathematics
Indiana University
Bloomington, IN 47405
U.S.A.

AMS Classifications: 10-01, 10-03, 10A15, 10A32, 10A45, 10C02, 10C05, 10C07, 10E20, 10E25, 10E35, 10G05, 10H05, 10H08, 10J05, 10L20, 12-03, 12A25, 12A50, 01A05, 01A45, 01A50, 01A55

Library of Congress Cataloging in Publication Data
Scharlau, Winfried.
    From Fermat to Minkowski.
    (Undergraduate texts in mathematics)
    Translation of: Von Fermat bis Minkowski.
    Bibliography: p.
    Includes index.
    1. Numbers, Theory of—History. I. Opolka, Hans.
II. Title. III. Series.
QA241.S2813 1984   512'.7'09   83-26216

This English edition is translated from *Von Fermat bis Minkowski: Eine Vorlesung über Zahlentheorie und ihre Entwicklung*, Springer-Verlag, 1980.

Printed in the United States of America.

9 8 7 6 5 4 3 2 1

ISBN 978-1-4419-2821-4

Preface

This book arose from a course of lectures given by the first author during the winter term 1977/1978 at the University of Münster (West Germany). The course was primarily addressed to future high school teachers of mathematics; it was not meant as a systematic introduction to number theory but rather as a historically motivated invitation to the subject, designed to interest the audience in number-theoretical questions and developments. This is also the objective of this book, which is certainly not meant to replace any of the existing excellent texts in number theory. Our selection of topics and examples tries to show how, in the historical development, the investigation of obvious or natural questions has led to more and more comprehensive and profound theories, how again and again, surprising connections between seemingly unrelated problems were discovered, and how the introduction of new methods and concepts led to the solution of hitherto unassailable questions. All this means that we do not present the student with polished proofs (which in turn are the fruit of a long historical development); rather, we try to show how these theorems are the necessary consequences of natural questions.

Two examples might illustrate our objectives. The book will be successful if the reader understands that the representation of natural numbers by quadratic forms—say, $n = x^2 + dy^2$—necessarily leads to quadratic reciprocity, or that Dirichlet, in his proof of the theorem on primes in arithmetical progression, simply had to find the analytical class number formula. This is why, despite some doubts, we retained the relatively amorphous, unsystematic and occasionally uneconomical structure of the original lectures in the book. A systematic presentation, with formal definitions, theorems, proofs and remarks would not have suited the real purpose of this course, the description of living developments. We nevertheless hope that the reader, with the occasional help of a supplementary text, will be able to learn a number of subjects from this book such as the theory of binary quadratic forms or of continued fractions or important facts on L-series and ζ-functions.

Clearly, we are primarily interested in number theory but we present it not as a streamlined ready-made theory but in its historical genesis, however, without inordinately many detours. We also believe that the lives and times of the mathematicians whose works we study are of intrinsic interest; to learn something about the lives of Euler and Gauss is a sensible supplement to learning mathematics. What was said above also applies to the history in this book: we do not aim at completeness but hope to stir up the interests of our readers by confining ourselves to a few themes and hope this will give enough motivation to study some of the literature quoted in our text.

Many persons have contributed to this book. First of all, the students of the course showed a lot of enthusiasm for the subject and made it worthwhile to prepare a set of notes; Walter K. Bühler kindly suggested to publish these notes in book form and prepared the English translation. Gary Cornell helped with the translation and suggested several mathematical improvements; many colleagues and friends contributed encouragement and mathematical and historical comments and pointed out a number of embarrassing errors. We wish to mention in particular Harold Edwards, Wulf-Dieter Geyer, Martin Kneser, and Olaf Neumann. It is a pleasure to thank them all.

Münster, West Germany   WINFRIED SCHARLAU
June 1984   HANS OPOLKA

Added in proof. In early 1984, André Weil's Number Theory: *An Approach Through History from Hammurapi to Legendre* appeared. It contains substantial additional material and discussion, especially concerning the period between Fermat and Legendre.

Contents

Literature

More references are given at the end of the individual chapters.

Original Sources

P. G. L. Dirichlet: *Works*, 2 vols., Reimer, Berlin, 1889, 1897 (reprinted by Chelsea, New York).
L. Euler: *Opera Omnia*, Series Prima, 23 vols., Teubner, Leipzig and Berlin, 1911–1938 (specifically Vols. 1, 2, 3 and 8)
P. de Fermat: *Ouevres*, 3 vols., Gauthier-Villars, Paris, 1891–1896 (specifically Vol. 2).
C. F. Gauss: *Werke*, 12 vols., Göttingen, 1870–1927 (specifically Vols. 1 and 2). *Disquisitiones Arithmeticae*, Yale University Press, New Haven, 1966.
J. L. Lagrange: *Ouevres*, 14 vols., Gauthier-Villars, Paris, 1867–1892 (specifically Vol. 3).
H. Minkowski: *Gesammelte Abhandlungen*, 2 vols., Teubner, Leipzig and Berlin, 1911 (reprint Chelsea, New York, 1967).

Textbooks on Number Theory

Z. I. Borewich, I. R. Shafarevich: *Number Theory*, Academic Press, New York, 1966.
P. G. L. Dirichlet: *Vorlesungen über Zahlentheorie, herausgegeben und mit Zusätzen versehen von R. Dedekind*, 4. Aufl., Vieweg und Sohn, Braunschweig, 1893 (reprint, Chelsea, New York, 1968).
H. M. Edwards: *Fermat's Last Theorem, A Genetic Introduction to Algebraic Number Theory*, Springer-Verlag, New York, Heidelberg, Berlin, 1977.
G. H. Hardy, E. M. Wright: *An Introduction to the Theory of Numbers*, Oxford University Press, New York, 1960.
H. Hasse: *Vorlesungen Über Zahlentheorie*, Springer-Verlag, Berlin, Göttingen, Heidelberg, 1950.
E. Hecke: *Lectures on the Theory of Algebraic Numbers*, Springer-Verlag, New York, 1981.

History of Mathematics

E. T. Bell: *Men of Mathematics*, Simon and Schuster, New York, 1957.

N. Bourbaki: *Elements d'histoire des Mathématiques*. Hermann, Paris, 1969.

M. Cantor: *Vorlesungen über Geschichte der Mathematik*, 4 Bände, Teubner Verlagsgesellschaft, Stuttgart, Neudruck, 1964.

L. E. Dickson: *History of the Theory of Numbers*, 3 vols., Carnegie Institute of Washington, 1919, 1920, and 1923 (reprint Chelsea, New York, 1971).

Ch. C. Gillispie (Editor): *Dictionary of Scientific Biography*, 15 vols., Scribner's, New York, 1970–1978.

F. Klein: *Vorlesungen über die Entwicklung der Mathematik*. im 19, Jahrhundert, 2 vols., Springer-Verlag, Berlin, 1926/27.

M. Kline: *Mathematical Thought from Ancient to Modern Times*, Oxford University Press, New York, 1972.

D. J. Struik: *A Concise History of Mathematics*, 3rd ed., Dover, New York, 1967.

A. Weil: Two lectures on number theory, past and present;
> Sur les sommes de trois et quatre carrés;
> La cyclotomie jadis et naguère;
> Fermat et l'équation de Pell;
> History of mathematics: why and how;

> all in *Collected Papers*, Vol. III, Springer-Verlag, New York, Heidelberg, Berlin, 1979.

CHAPTER 1
The Beginnings

The first work devoted to topics from the history of mathematics of which at least a few fragments survive is by the Greek mathematician Eudemus of Rhodes, a member of the school of Aristotle. It begins as follows (quoted from *Proclus:*

> Next we must speak of the development of this science during the present era . . . we say, as have most writers of history, that geometry was first discovered among the Egyptians and originated in the remeasuring of their lands. This was necessary for them because the Nile overflows and obliterates the boundary lines between their properties. It is not surprising that the discovery of this and the other sciences had its origin in necessity, since everything in the world of generation proceeds from imperfection to perfection. Thus they would naturally pass from *sense-perception* to *calculation* and from *calculation* to *reason*. Just as among the Phoenicians the necessities of trade and exchange gave the impetus to the accurate study of number, so also among the Egyptians the invention of geometry came about from the cause mentioned.

According to Eudemus the Phoenicians invented number theory but this is probably false; rather, we should look to Babylon for the origins of arithmetic and algebra. It is more interesting to see what Eudemus' main message is: If we interpret his first sentence in a general way, we can understand it as motivating the necessity of studying history. The followers of Aristotle liked to assign an author to every idea, a tendency which makes Eudemus' statement not very surprising. Nonetheless, the most eminent mathematicians have emphasized again and again how important it was for them to go back to study the original papers of their predecessors. This is often connected with a direct encouragement to the reader to follow the masters' footsteps. Today, where one never learns mathematics from the originals but rather from secondary sources, this is certainly not clear. However, one of the main objectives of this book is to guide the reader to the study of the works of some of the greatest number theorists.

Let us stop for a second and consider the statement that science proceeds from the imperfect to the perfect. This appears to be a triviality, and mathematicians will certainly agree with it. But it gains a new dimension when we keep in mind that generally the history of mankind does not seem to proceed or progress from the imperfect to the perfect. Some of our readers might feel challenged to reflect on whether this is a real or only an apparent contradiction. To us, it is particularly important to see how Eudemus describes the development of a mathematical theory, and this is a legitimate reading of what he says. In mathematics, *sense perception* may entail an interesting numerical example, or a specific problem that we want to solve. *Calculation* tries to solve the problem in a more general framework, perhaps by determining all solutions of an equation or by finding necessary and sufficient conditions for its solvability. And *reason* is the imbedding of a specific problem in a more general theory, the generalization of special cases, or the search for the "real reasons." We want to illustrate this with the help of an example: the only number-theoretical problem that we believe to have been fully solved in antiquity.

Let us take as a "sense-perception" the equation $3^2 + 4^2 = 5^2$ which has a well-known geometrical interpretation by the Pythagorean theorem—which was known long before Pythagoras (approximately 580–500 B.C.). (If this does not suffice as "sense perception" one can convince oneself that a string, stretched in a corresponding right triangle, is tuned in the proportion *keynote* : *quart* : *sext*.) Other Pythagorean triples have been known for a long time and in many different cultures:

$$5^2 + 12^2 = 13^2,$$
$$7^2 + 24^2 = 25^2,$$
$$8^2 + 15^2 = 17^2.$$

Now it is not too far-fetched (?!) to try to determine all such triples. The first observation is that one can generate new triples by multiplying one of the equations by a square. One can invert this idea and cancel, as far as possible, square factors. This means that it is enough to investigate the equation

$$a^2 + b^2 = c^2$$

only when a, b, and c have no common factor. Specifically, all three numbers are not even, neither are all of them odd. (Now we are in the midst of what Eudemus called "calculation"!) Exactly one of the three numbers is even. Which one? Giving the matter some more thought we see that it cannot be c. For then c^2 would be a multiple of 4 while since a and b

are both odd, say $a = 2d + 1$ and $b = 2f + 1$, $a^2 + b^2$ could be written as

$$a^2 + b^2 = 4(d^2 + d + f^2 + f) + 2$$

which is not a multiple of 4. Let a now denote the number that is even in our Pythagorean triple. We transform the equation into

$$a^2 = c^2 - b^2 = (c - b)(c + b).$$

All factors are positive even numbers, so we set

$$a = 2n, \qquad c - b = 2v, \qquad c + b = 2w$$

which yields $n^2 = vw$. What happened along the way to our condition that a, b, and c are relatively prime? Answer: v and w are relatively prime and not both odd because otherwise $b = w - v$, $c = w + v$ would have a common factor. Then, because of unique prime factorization, a theorem already known to Euclid (approx. 300 B.C.), the factors v and w in the equation $n^2 = vw$ have to be perfect squares because v and w are relatively prime.

Now turn this around: Let v and w be arbitrary relatively prime squares of different parity, say $v = p^2$ and $w = q^2$ with $q > p$. If we set

$$a = 2pq, \qquad b = q^2 - p^2, \qquad c = q^2 + p^2,$$

then we see that every Pythagorean triple must be of this form. Conversely, a, b, and c are, as one can easily see, relatively prime and

$$a^2 + b^2 = (2pq)^2 + (q^2 - p^2)^2 = 4p^2q^2 + q^4 - 2p^2q^2 + p^4$$

$$= q^4 + 2p^2q^2 + p^4 = (q^2 + p^2)^2 = c^2.$$

The Babylonian mathematician (unfortunately we do not know his name) who apparently knew all this some 3500 years ago did what we would do: he asked one of his assistants to compute a list of the first 60, 120, or 3600 Pythagorean triples and to write them down. He used clay tables; we would write a small program for our pocket calculator and print out the results. If we could ask the Babylonian mathematician why he wanted to know these numbers he would perhaps give us answers as obscure as those we would give and certainly not as clear as

$$4961^2 + 6480^2 = 8161^2.$$

We have already mentioned that these considerations mark the highest achievements of ancient number theory. Whatever was known beyond this can be described in a few sentences: the most important divisibility theorems for integers including the Euclidean algorithm and unique factorization into primes, the summation of simple finite series such as

$$1 + 2 + 3 + \cdots + n = \frac{n(n + 1)}{2}$$

or

$$1 + k + k^2 + \cdots + k^n = \frac{k^{n+1} - 1}{k - 1}$$

and, as perhaps the most remarkable single result, Euclid's theorem that for any prime of the form $1 + 2 + \cdots + 2^n = 2^{n+1} - 1 = p$, the number $2^n p$ is a perfect number, i.e., it is the sum of its proper divisors. (In this connection, the following remark: to determine all perfect numbers is the oldest unsolved problem in number theory, probably the oldest unsolved mathematical problem. Euler showed that any even perfect number has the form given by Euclid. No odd perfect number is known, but their nonexistence has not been proved.)

The other number-theoretical results that were known in antiquity concern not general problems but rather special numerical equations or systems of equations. Many of those results, often with tricky solutions, can be found in the works of Diophantos (approximately 250 A.D.). Contrary to modern terminology where "Diophantine solutions" are always integers, Diophantos himself allows rational solutions. This means that his work is algebraic rather than number theoretic. Of course, this distinction is superficial; just consider the above equation, $a^2 + b^2 = c^2$. Having a rational solution, one obtains an integral solution by multiplication by a common denominator; conversely, one obtains rational solutions from integral solutions by dividing by arbitrary squares of integers. In fact, Diophantos' derivation is quite similar to our solution above. Moreover, Diophantos knew some basic theorems about the representations of numbers as sums of squares but largely without proofs or sometimes with only partial proofs. His work was an important source of inspiration for later mathematicians, particularly Fermat. One can agree completely with Jacobi's words: "Diophantos will always be remembered because he started the investigation of the deep-rooted properties and relations between numbers which have been understood by the beautiful research of modern mathematics."

References

Proclus, *A Commentary on the First Book of Euclid's Elements*, translated with Introduction and Notes by Glenn R. Morrow, Princeton University Press, Princeton, N.J., 1970.

Edwards, Chap. 1.

Th. L. Heath: *Diophantus of Alexandria*, Dover, New York, 1964.

CHAPTER 2
Fermat

After more than a thousand years of stagnation and decay the rejuvenation and revitalization of western mathematics, particularly algebra and number theory, starts with Leonardo of Pisa, known as Fibonacci (ca. 1180–1250). Occasionally, the formula

$$(a^2 + b^2)(c^2 + d^2) = (ac - bd)^2 + (ad + bc)^2 \qquad (2.1)$$

is ascribed to Fibonacci: if two numbers are the sums of two squares, their product is a sum of two squares as well. This development was continued by the Italian renaissance mathematicians Scipio del Ferro (ca. 1465–1526), Nicolo Fontano, known as Tartaglia (ca. 1500–1557), Geronimo Cardano (1501–1576), and Ludovico Ferrari (1522–1565). Their solution of algebraic equations of the third and fourth degree marks the first real progress over ancient mathematics. Next in this line is François Viète (1530–1603) who introduced the use of letters in mathematics. With Viète, we enter the seventeenth century; from that time on, mathematics enjoys an uninterrupted, continuous and exponentionally accelerating development. This new era, the era of modern mathematics, starts with four great French mathematicians: Girard Desargues (1591–1661), René Descartes (1596–1650), Pierre de Fermat (1601–1665), and Blaise Pascal (1623–1662).

It is difficult to imagine four persons more different from each other: Desargues—the most original, an architect by profession—was thought of as strange, an individualist who wrote his main opus in a kind of secret code and had it printed in minute letters. Descartes—the most famous—started out as a professional soldier and was well able to fend off a group of sailors who wanted to rob him. Like a professional soldier, he planned a general attack (*Discours sur la méthode*) on the foundations of science. Pascal—the most ingenious—left mathematics and became a religious

fanatic, troubled by constipation for most of his short life. Finally, Fermat
—the most important—was royal councillor at the Parliament of Toulouse,
a position that, in today's terms, can be described as a high-level adminis-
trator.

Fermat's profession apparently provided him with all the leisure he
needed to occupy himself with mathematics. His style of work was slow, his
letters, which contain all his important number-theoretical results, are
laconic and dry. The majority of these were directed to Mersenne who, for
a while, was Fermat's go-between in exchanges with other mathematicians.
Several of these correspondents were important in the development of
number theory, among them Frenicle, Pascal, and Carcavi. In these letters,
Fermat formulated number-theoretical problems, but there are also several
definitive statements and discussions of special numerical examples.

Fermat never gave proofs and only once did he indicate his method of
proof. (We will come back to this later.) This makes it difficult to determine
what Fermat really proved as opposed to what he conjectured on the basis
of partial results or numerical evidence. We will see that many of his
theorems cannot be proved easily, and first-rate mathematicians, such as
Euler, had great trouble proving them. On the other hand, there can be no
doubt that Fermat knew how to prove many if not most of his theorems
completely. His letters indicate that at about 1635, inspired by Mersenne,
Fermat began to occupy himself with number-theoretical questions. His
first interests were perfect numbers, amicable numbers, and similar arithmetical brain-teasers. He describes several ways to construct such numbers,
but far more remarkable is that—showing more insight than any of his
contemporaries—he succeeded in proving an important theorem in this still
very barren area, "Fermat's little theorem": $a^{p-1} \equiv 1 \mod p$ for every
prime number p and every number a prime to p. (Today this theorem is
proved early in an algebra course from the basic notions of group theory.)
Fermat's most important number-theoretical heritage is a letter to Carcavi
in August 1650 (Fermat, *Oeuvres*, II, pp. 431–436). He himself considers
this as his testament, a fact which he expresses in the following words:
"Voilà sommairement le compte de mes reveries sur le sujet des nombres."
At the beginning of this letter, one finds the passage where he describes a
certain method of proof which he himself discovered and used with great
success. He then formulates a number of theorems all of which were
contained in earlier letters or papers, but it is obvious that he wanted to
compile what he himself considered his most beautiful and important
results.

Fermat writes the following about his method of proof, quoted from
E. T. Bell, *Men of Mathematics*:

> For a long time I was unable to apply my method to affirmative propositions,
> because the twist and the trick for getting at them is much more troublesome
> than that which I use for negative propositions. Thus, when I had to prove
> that *every prime number which exceeds a multiple of* 4 *by* 1 *is composed of two*

squares, I found myself in a fine torment. But at last a meditation many times repeated gave me the light I lacked, and now affirmative propositions submit to my method, with the aid of certain new principles which necessarily must be adjoined to it. The course of my reasoning in affirmative propositions is such: if an arbitrarily chosen prime of the form $4n + 1$ is not a sum of two squares, [I prove that] there will be another of the same nature, less than the one chosen, and [therefore] next a third still less, and so on. Making an infinite descent in this way we finally arrive at the number 5, the least of all the numbers of this kind $[4n + 1]$. [By the proof mentioned and the preceding argument from it], it follows that 5 is not a sum of two squares. But it is. Therefore we must infer by a *reductio ad absurdum* that all numbers of the form $4n + 1$ are sums of two squares.

This method is now called *the method of infinite descent*. Before going into more details we want to give a brief explanation of what Fermat might have been thinking when he spoke of the relative simplicity with which one can prove negative statements. First we consider an example using the same principle encountered in our earlier discussion of Pythagorean triples.

(2.2) **Theorem.** *No natural number of the form* $8n + 7$ *is the sum of three squares.*

PROOF. Let k be a natural number (including 0). If one divides k^2 by 8, the remainder will be 0, 1, or 4: When k is even, the remainder will be either 0 or 4, for odd $k = 2l + 1$, the remainder is always 1 because $k^2 = 4(l^2 + l) + 1$. Consequently, after forming the sum of three squares of natural numbers, division by 8 leaves a remainder $p + q + r$ with p, q, r either 0, 1, or 4. Checking all the possibilities shows that the remainder will be 0, 1, 2, 3, 4, 5, or 6 but not 7.

It is easy to obtain numerous similar negative results. For example, try to determine which numbers cannot be written as sums of two squares.

We now formulate most of the theorems listed in Fermat's letter to Carcavi.

(2.3) **Theorem** (Two-Square Theorem). *Every prime number of the form* $4k + 1$ *can be written uniquely as a sum of two squares.*

(2.4) **Theorem** (Four-Square Theorem). *Every natural number is a sum of four squares of natural numbers* (*zero is allowed as a summand*).

(2.5) **Theorem.** *Let* N *not be a square. Then the equation*

$$Nx^2 + 1 = y^2$$

has infinitely many integer solutions.

Pierre de Fermat

This equation is frequently called Pell's equation [because the English mathematician Pell (1610–1685) had nothing to do with it].

(2.6) **Theorem.** *The equation*

$$x^3 + y^3 = z^3$$

cannot be solved in natural numbers.

(2.7) **Theorem.** *The only solution in natural numbers of the equation*

$$x^3 = y^2 + 2$$

is $y = 5$, $x = 3$.

Finally, there is an assertion that was later disproved by Euler; namely:

(2.8) *Every integer of the form $2^{2^n} + 1$ is prime.*

Euler gave the example

$$2^{2^5} + 1 = 4294967297 = 641 \cdot 6700417.$$

Quite correctly, Fermat writes in his letter, "Il y a infinies questions de cette espèce, . . . ," and it is remarkable with what certainty he identified central problems in number theory. Each of the theorems we have just listed is the starting point for a deep and rich theory. This is true even for the incorrect assertion (2.8). The so-called Fermat numbers $2^{2^n} + 1$ occur in Gauss' solution to the problem of constructing the regular k-gon. According to Gauss, the division of the circle into k parts with ruler and compass is possible for odd k only if k is a square free product of Fermat primes. Jacobi writes the following about Fermat's statements concerning the "quadratic forms" $x^2 + y^2, x^2 + 2y^2, x^2 + 3y^2, x^2 - dy^2, \ldots$ (in his *Collected Papers*, Vol. 7): "The efforts of mathematicians to prove these theorems have created the great arithmetical theory of quadratic forms." There is nothing we have to add to Descartes' statement about the four-square theorem: "Without any doubt this theorem is one of the most beautiful that can be found in number theory but I do not know a proof; in my judgement, it will be so difficult that I did not even attempt to search for it." Euler, the most important mathematician of the eighteenth century, tried for 40 years to find a proof, without succeeding. We know from the correspondence with Mersenne that Fermat knew most of these theorems before 1638.

Before starting to explain some of the proofs, we want to give a better idea of the development of number theory by the middle of the seventeenth century by quoting a few further statements and propositions from Fermat's correspondence.

(2.9) **Theorem.** *The equation $x^4 + y^4 = z^2$, and more specifically the equation $x^4 + y^4 = z^4$, is not solvable in integers.*

(2.10) **Theorem** (from a letter to Pascal on September 25, 1654). *Every prime number of the form $3k + 1$ can be written as $x^2 + 3y^2$. Every prime number of the form $8k + 1$ or $8k + 3$ can be written as $x^2 + 2y^2$.*

(2.11) **Theorem.** *Every number is the sum of at most three triangular numbers, i.e., numbers of the form $n(n-1)/2 = \binom{n}{2}$.*

Approximately 150 years later, Gauss proved that every natural number which is not of the form $4^k(8n + 7)$ can be written as the sum of three squares. This result is basically equivalent to the theorem on triangular numbers.

(2.12) **Theorem.** *No triangular number (with the exception of* 1) *is a cube.*

Though we could add many more interesting results, let us stop here. It is obvious that the majority of these theorems remain central to lectures or books on elementary number theory even today.

We will now give a detailed proof of (2.9). There are three reasons why we start with this theorem. It is the only theorem for which Fermat himself published a rather complete proof. Also, proving it, we can continue where we left off when determining the Pythagorean triples in Chapter 1. Finally, this is the simplest result to which we may apply the method of infinite descent.

PROOF OF (2.9). Let us assume that the proposition is wrong. Then there are pairwise relatively prime integers x, y, z with $x^4 + y^4 = z^2$. We assume that z is minimal with this property. It follows from our study of Pythagorean triples (Chapter 1) that $x^2 = A^2 - B^2$, $y^2 = 2AB$, $z = A^2 + B^2$, where B is even. Since x and y are relatively prime, A and B are as well, and therefore $A = a^2$, $B = 2b^2$. Consequently, $x^2 + (2b^2)^2 = a^4$ which leads to $2b^2 = 2CD$, $a^2 = C^2 + D^2$. The numbers C and D are relatively prime. One obtains $C = c^2$, $D = d^2$; consequently $a^2 = c^4 + d^4$. Now $z = a^4 + (2b^2)^2 > a^4 \geqslant a$, and we know $b > 0$ because $y > 0$. This is a contradiction to the fact that z is minimal.

Before proving the two-square theorem (also with the help of the method of infinite descent), we prove the following lemma:

(2.13) **Lemma.** *The equation $x^2 + y^2 = 0$ has a nontrivial solution in the field $\mathbb{F}_q$ of q elements if and only if q is of the form $q = 4k + 1$.*

PROOF. For $q = 4k + 1$ the multiplicative group $\mathbb{F}_q^*$ of $\mathbb{F}_q$ contains an element x of order 4, i.e., $x^2 + 1 = x^2 + 1^2 = 0$ in $\mathbb{F}_q$. Let us assume a nontrivial solution (x, y) exists for $q = 4k + 3$. Then x/y is a nontrivial element in $\mathbb{F}_q^*$ of order 4. This is a contradiction because the order of the group $\mathbb{F}_q^*$ is $q - 1 = 4k + 2 = 2(2k + 1)$.

It is, we think, reasonable to assume that Fermat thought of a proof of the two-square theorem along the following lines.

PROOF OF (2.3) First, let us prove existence. Assume that the statement is false. Then there is a minimal prime number p with $p = 4k + 1$, which is not the sum of two integer squares. Let n be the smallest natural number such that $np = x^2 + y^2$ with x, y relatively prime. Such an n exists because by (2.13) we know $mp = z^2 + u^2$ for suitable integers. We can assume that $x, y > 0$ and that $x, y < p/2$. This can be easily proved by determining integers k, l with $|x - kp| < p/2$ and $|y - lp| < p/2$ and replacing x, y by $|x - kp|$ and $|y - lp|$. Another consequence is $n < p/2$. Moreover x and y are both relatively prime. Otherwise there would be a prime number q with $x = x_0 q$, $y = y_0 q$ and consequently $np = q^2(x_0^2 + y_0^2)$. Then, $q < p$ and consequently q^2 is a divisor of n because $x, y < p/2$. This means that n/q^2 is a natural number n, and consequently $(n/q^2)p$ is the sum of two squares, contradicting our choice of n. Specifically, x, y are not both even nor are they both odd, for if they were, 2 would be a divisor of both $x \pm y$ and n, and we would obtain the following contradiction: $(n/2)p = (x + y)^2/4 + (x - y)^2/4$. To complete this proof, we observe that n contains only prime factors of the form $4k + 1$, for if there were a prime factor of the form $q = 4k + 3$ then, according to (2.13), x and y would have q as a common factor. This contradicts the fact that x and y are relatively prime. Let q be a prime factor of n. Since $n < p$, we have $q < p$, and q can be written as a sum of two squares: $q = u^2 + v^2$. Then

$$\frac{n}{q} p = \frac{1}{q}(x^2 + y^2) = \frac{x^2 + y^2}{u^2 + v^2}$$

$$= \left(\frac{ux + vy}{u^2 + v^2}\right)^2 + \left(\frac{uy - vx}{u^2 + v^2}\right)^2 \tag{*}$$

$$= \left(\frac{ux - vy}{u^2 + v^2}\right)^2 + \left(\frac{uy + vx}{u^2 + v^2}\right)^2. \tag{**}$$

In the field with q elements one has: $v^2/u^2 = -1$ and $x^2/y^2 = -1$; consequently $v/u = \pm x/y$ or $vy \pm xu = 0$. This means that (*) or (**) provides a representation of $(n/q)p$ as a sum of two squares of integers, which contradicts our assumption.

Now, we prove uniqueness. Let us assume that

$$p = x^2 + y^2 = X^2 + Y^2. \tag{+}$$

There are exactly two solutions of the congruence $z^2 + 1 \equiv 0 \bmod p$ [c.f. (2.13)]. They can be written as $z \equiv \pm h \bmod p$. Consequently, $x \equiv \pm hy$ mod p and $X \equiv \pm hY \bmod p$. Since the sign does not matter, we choose

$$x \equiv hy \bmod p, \qquad X \equiv -hY \bmod p. \tag{++}$$

From (+) and Fibonacci's formula (2.1) we have

$$p^2 = (x^2 + y^2)(X^2 + Y^2) = (xX - yY)^2 + (xY + yX)^2.$$

Because of (++), $xY + yX \equiv 0 \bmod p$ and consequently $xX - yY \equiv 0$ mod p. Division by p^2 yields a representation of 1 as a sum of two integral squares. The only possible representation of this kind is $1 = (\pm 1)^2 + 0^2$. This shows $xX - yY$ or $xY + yX = 0$. Uniqueness follows from the fact that x, y, X, Y are pairwise relatively prime.

From what we have proved so far and Fibonacci's formula (2.1), it follows that all numbers divisible only by 2 and numbers of the form $4k + 1$ can be written as sums of two squares. Once one knows that every prime number p of the form $4k + 1$ can be written uniquely in the form $x^2 + y^2$, it is natural to ask how to construct such numbers x, y. Here we mention only that there are several known methods [Legendre (1808), Gauss (1825), Serret (1848), and Jacobsthal (1906)]. Lengendre's method is based on the theory of continued fractions which will be discussed in Chapter 5 and used in our proof of the important Theorem (2.5). Let us now make a few comments on the equation $x^2 - dy^2 = 1$; (this is now the usual way of writing the equation which occurs in (2.5)). One reason why this equation is so interesting is that using it one can find an optimal rational approximation of $\sqrt{d}$. For large x, y, one has $\sqrt{d} \approx x/y$ if $x^2 - y^2 d = 1$. However, it is mathematically much more interesting that the smallest solutions of the equation do not appear to follow any regular pattern. This can be seen in Table 1 which lists the smallest solutions for a few d. Fermat seems to have computed a similar table because in his letters he poses the equation $x^2 - y^2 d = 1$ as a problem and repeatedly chooses special d for which x and y become particularly large, e.g., $d = 61, 109, 149$. (Centuries earlier, Indian mathematicians seem to have known much about this equation.)

Table of the smallest solutions of $x^2 - dy^2 = 1$.

d	x	y
8	3	1
10	19	6
11	10	3
12	7	2
13	649	180
14	15	4
15	4	1
60	31	4
61	1766319049	226153980
62	63	8
108	1351	130
109	158070671986249	15140424455100
110	21	2
148	73	6
149	25801741449	2113761020
150	49	4

And who will be able to see right away that with $d = 991$

$$x = 379516400906811930638014896080,$$
$$y = 12055735790331359447442538767 ?$$

To conclude this chapter, a few words about the so-called "Fermat's last theorem." Whether Fermat knew a proof or not has been the subject of many speculations. The truth seems to be obvious. Fermat made his famous remark in the margin of his private copy of Bachet's edition of Diophantos in 1637 (next to the problem of decomposing a square into the sum of two squares): "Cubum autem in duos cubos, aut quadrato-quadratum in duos quadrato-quadratos, et generaliter nullam in infinitum ultra quadratum potestatem in duas ejusdem nominis fas est dividere: cujus rei demonstrationem mirabilem sane detexi. Hanc marginis exiguitas non caperet." (Fermat, *Oeuvres*, III, p. 241). Basically, Fermat claims that the equation $x^n + y^n = z^n$, $n \geqslant 3$, is unsolvable in natural numbers and states that he has a truly wonderful proof. The margin, however, was too small to write it down.

This statement was made at the time of his first letters concerning number theory and we can assume that this was also the time his interest awakened in the theory of numbers. As far as we know, he never repeated his general remark but repeatedly made the statement for the cases $n = 3$ and 4 and posed these cases as problems to his correspondents. We have already seen that he formulated the case $n = 3$ in a letter to Carcavi in 1659 ($n = 4$ obviously appeared to be too simple to be included in his collection of important theorems). All these facts indicate that Fermat quickly became aware of the incompleteness of the "proof" of 1637. Of course, there was no reason for a public retraction of his privately made conjecture.

References

Fermat, *Oeuvres*, Vol. 2.
Edwards, Chap. 1.
Bell, Chap. 4.
Th. L. Heath (cf. Literature to Chapter 1).
J. E. Hofmann: Fermat, Pierre de (in Dictionary of Scientific Biography).
J. E. Hofmann: Über zahlentheoretische Methoden Fermats and Eulers, ihre Zusammenhänge und Bedeutung, Arch. History Exact Sciences 1, (1961), 122–159.
M. S. Mahoney: *The Mathematical Career of Pierre de Fermat (1601–1665)*, Princeton University Press, Princeton, NJ, 1973.

Also:
A. Weil: Review of Mahoney's book in *Works* Vol. III.
L. J. Mordell: *Three Lectures on Fermat's Last Theorem*, Cambridge University Press, Cambridge, England, 1921. (Reprint: Chelsea.)
P. Bachmann: *Das Fermatproblem in seiner bisherigen Entwicklung*, Springer-Verlag, Berlin-Heidelberg-New York, Neudruck, 1976.
H. Davenport: *The Higher Arithmetic*, third edition, Hutchinson, 1960; fifth edition, Cambridge University Press, Cambridge, England, 1982.

CHAPTER 3
Euler

After 1650 number theory stood virtually still for a hundred years. This period saw the development of analysis in the work of Isaac Newton (1643–1727), Gottfried Wilhelm Leibniz (1646–1716), the Bernoullis (Jacob, 1655–1705; Johann I, 1667–1748; Nicholas II, 1687–1759; Daniel 1700–1792), and Leonhard Euler (1707–1783). Analysis is not the subject of this book, but analytic methods have played an important role in number theory since Dirichlet. This interplay between analysis and number theory has its origins in the work of Euler, and we will try to sketch the beginnings of this development here.

One of the most important and interesting objects in analysis was the geometrical series which was first summed by Nicole Oresme (ca. 1323–1382):

$$1 + x + x^2 + \cdots = \frac{1}{1-x} \qquad \text{for} \quad |x| < 1.$$

Comparatively simple manipulations, well known since the beginnings of analysis, lead to other series, e.g.,

$$\frac{1}{1+x} = 1 - x + x^2 - x^3 + \cdots ;$$

by integration we get

$$\log(1 + x) = x - \frac{x^2}{2} + \frac{x^3}{3} - + \cdots .$$

Applying Abel's limit theorem yields

$$\log(2) = 1 - \tfrac{1}{2} + \tfrac{1}{3} - + \cdots .$$

Another example is

$$\frac{1}{1+x^2} = 1 - x^2 + x^4 - x^6 + - \cdots ;$$

by integration one obtains

$$\arctan x = x - \frac{x^3}{3} + \frac{x^5}{5} - + \cdots ,$$

and with the help of Abel's limit theorem,

$$\frac{\pi}{4} = \arctan 1 = 1 - \tfrac{1}{3} + \tfrac{1}{5} - \tfrac{1}{7} + - \cdots .$$

To this last series, well known to the reader from lectures in analysis, we will return repeatedly. Its discovery, as the discovery of many fundamental results in analysis, can be ascribed to several mathematicians. Gregory seems to have been the first, but Leibniz independently found it approximately 1663, i.e., before (re)discovering the fundamental theorem of calculus. It is quite possible that this immediately acknowledged outstanding achievement (by, among others, Huygens) prompted Leibniz, who was then a lawyer and diplomat, to turn to mathematics.

These series show—though in a naive and superficial way—that there is a connection between sequences of integers satisfying a simple mathematical principle and transcendental functions. We will encounter such examples again and again, and we now introduce a particularly important one, related to the so-called Bernoulli numbers which are of great importance in mathematics even today.

The function

$$f(x) = \frac{x}{e^x - 1} ,$$

has a convergent power series expansion in the unit disc,

$$f(x) = B_0 + \frac{B_1}{1!} x + \frac{B_2}{2!} x^2 + \cdots + \frac{B_n}{n!} x^n + \cdots .$$

The coefficients B_i are called the "Bernoulli numbers" (after Jacob Bernoulli). We will now give a recursive formula for these numbers.

$$\left(\sum_{m=0}^{\infty} a_m x^m \right) \left(\sum_{n=0}^{\infty} b_n x^n \right) = \sum_{r=0}^{\infty} \left(\sum_{k=0}^{r} a_k b_{r-k} \right) x^r$$

is the Cauchy product of two power series. If the right-hand side of this formula equals 1, one obtains

$$a_0 b_0 = 1, \qquad a_0 b_n = - \sum_{k=0}^{n-1} b_k a_{n-k} .$$

This permits us to compute b_n recursively. Because

$$\frac{e^x - 1}{x} = \sum_{n=0}^{\infty} \frac{1}{(n+1)!} x^n,$$

we obtain the following recursive formula for the Bernoulli numbers:

$$B_n = -\sum_{k=0}^{n-1} B_k \frac{1}{k!} \frac{n!}{(n-k+1)!}. \tag{3.1}$$

According to (3.1) all the B_n are rational numbers. We have

$$B_0 = 1, \qquad B_1 = -\tfrac{1}{2}, \qquad B_2 = \tfrac{1}{6}, \qquad B_3 = 0, \qquad B_4 = -\tfrac{1}{30},$$

$$B_5 = 0, \qquad B_6 = \tfrac{1}{42}, \qquad B_7 = 0, \qquad B_8 = -\tfrac{1}{30}, \qquad B_9 = 0,$$

$$B_{10} = \tfrac{5}{66}, \qquad B_{11} = 0, \qquad B_{12} = -\tfrac{691}{2730}, \cdots.$$

Euler computed B_k for $k \leqslant 30$. The B_k seem to occur first in Jacob Bernoulli's *Ars conjectandi*, Basel, 1713, in connection with computing the sum $\Sigma_\nu \nu^k$, $1 \leqslant \nu \leqslant n-1$. The computation of such sums was of interest even before Fermat, and Fermat himself did some work on them. Today, Bernoulli numbers appear in many places in number theory, but also in other areas such as algebraic topology. One feels that they are connected with particularly deep and central questions. Let us discuss some of the more elementary properties of these numbers.

First, let us rewrite the recursive formula:

$$\sum_{k=0}^{n} \frac{n!}{k!(n-k+1)!} B_k = 0.$$

Multiplying by $n+1$ and using the identity

$$\frac{(n+1)!}{k!(n-k+1)!} = \binom{n+1}{k}$$

yields

$$\sum_{k=0}^{n} \binom{n+1}{k} B_k = 0.$$

Using for $p(x) = \sum_{k=0}^{n} a_k x^k$ the notation

$$p(B) := \sum_{k=0}^{n} a_k B_k,$$

the recursion formula can be written very concisely as

$$(1 + B)^{n+1} - B^{n+1} = 0 \tag{3.1'}$$

The above table suggests the following.

(3.2) Remark. $B_k = 0$ for odd $k > 1$.

PROOF. Obviously, this statement is equivalent to the statement that $x/(e^x - 1) + x/2$ is an even function. This is easy to prove since

$$\frac{x}{e^x - 1} + \frac{1}{2}x = \frac{-x}{e^{-x} - 1} - \frac{1}{2}x$$

is equivalent to

$$\frac{x}{e^x - 1} + \frac{x}{e^{-x} - 1} = -x,$$

which is true because $e^x e^{-x} = 1$.

We mentioned above that Bernoulli numbers occurred for the first time in the formula

$$1^k + 2^k + \cdots + (n-1)^k = \frac{1}{k+1}\left((n+B)^{k+1} - B^{k+1}\right); \qquad (3.3)$$

specifically, for $k = 1, 2, 3$:

$$1 + 2 + \cdots + (n-1) = \tfrac{1}{2}(n-1)n,$$

$$1 + 2^2 + \cdots + (n-1)^2 = \tfrac{1}{6}n(n-1)(2n-1),$$

$$1 + 2^3 + \cdots + (n-1)^3 = \tfrac{1}{4}n^2(n-1)^2.$$

PROOF. Obviously,

$$\frac{e^{nx} - 1}{x}\frac{x}{e^x - 1} = \sum_{r=0}^{n-1} e^{rx} = \sum_{r=0}^{n-1}\sum_{k=0}^{\infty} \frac{r^k x^k}{k!} = \sum_{k=0}^{\infty}\left(\sum_{r=0}^{n-1} \frac{r^k}{k!}\right)x^k.$$

On the other hand, using the Cauchy product with $t = k - s$,

$$\frac{e^{nx} - 1}{x}\frac{x}{e^x - 1} = \left(\sum_{s=0}^{\infty} \frac{n^{s+1}x^s}{(s+1)!}\right)\left(\sum_{t=0}^{\infty} \frac{B_t}{t!}x^t\right)$$

$$= \sum_{k=0}^{\infty}\left(\sum_{s=0}^{k} \frac{n^{s+1}B_{k-s}}{(s+1)!(k-s)!}\right)x^k.$$

Comparing coefficients yields

$$\sum_{r=0}^{n-1} \frac{r^k}{k!} = \sum_{s=0}^{k} \frac{n^{s+1}B_{k-s}}{(s+1)!(k-s)!}.$$

and multiplication by $k!$

$$\sum_{r=0}^{n-1} r^k = \frac{1}{k+1}\sum_{s=0}^{k} \frac{(k+1)!}{(s+1)!(k-s)!}n^{s+1}B_{k-s}$$

$$= \frac{1}{k+1}\sum_{s=0}^{k}\binom{k+1}{s+1}n^{s+1}B_{k-s}$$

$$= \frac{1}{k+1}\left((n+B)^{k+1} - B^{k+1}\right).$$

We now come to one of Euler's most celebrated theorems, discovered in 1736 (*Institutiones Calculi Differentialis*, Opera (1), Vol. 10).

(3.4) Theorem.

$$\sum_{n=1}^{\infty} \frac{1}{n^{2k}} = \frac{2^{2k-1}\pi^{2k}|B_{2k}|}{(2k)!}.$$

Specifically, for $k = 1, 2,$ and 3 one obtains

$$\sum_{n=1}^{\infty} \frac{1}{n^2} = \frac{\pi^2}{6}, \qquad \sum_{n=1}^{\infty} \frac{1}{n^4} = \frac{\pi^4}{90}, \qquad \sum_{n=1}^{\infty} \frac{1}{n^6} = \frac{\pi^6}{945}.$$

The series $\sum_{n=1}^{\infty}(1/n^{2k})$ converges because $1/n^{2k} \leqslant 1/n^2 < 1/n(n-1)$ for $n \geqslant 2$ and

$$\sum_{n=2}^{k} \frac{1}{n(n-1)} = \sum_{n=2}^{k}\left(\frac{1}{n-1} - \frac{1}{n}\right) = 1 - \frac{1}{k}.$$

For a proof of (3.4) we need a few facts from the theory of trigonometric functions. First we make sort of a pedagogical remark: the usual definitions of the sine and cosine functions,

$$\sin x = \sum_{n=0}^{\infty} (-1)^n \frac{x^{2n+1}}{(2n+1)!},$$

$$\cos x = \sum_{n=0}^{\infty} (-1)^n \frac{x^{2n}}{(2n)!}$$

do not make it obvious that these functions are periodic. It is natural to look for an expression that makes the periodicity obvious. The simplest approach is to consider

$$f(x) = \sum_{n=-\infty}^{\infty} \frac{1}{x+n}. \tag{3.5}$$

Let us assume for a moment that this definition makes sense. Obviously, the right-hand side has period 1, for if one replaces x by $x + 1$, one just replaces the summation index n by $n + 1$. Now let us rewrite (3.5):

$$f(x) := \frac{1}{x} + \sum_{n=1}^{\infty} \left(\frac{1}{x+n} + \frac{1}{x-n}\right) = \frac{1}{x} + \sum_{n=1}^{\infty} \frac{2x}{x^2 - n^2}.$$

Clearly, $f(x)$ is not defined for $x \in \mathbb{Z}$, but it is easy to show that $\sum_{n=1}^{\infty}(1/(x+n) + 1/(x-n))$ is absolutely convergent for $x \notin \mathbb{Z}$. How can we express f by a known periodic function? The answer is easy:

$$f(x) = \frac{1}{x} + \sum_{n=1}^{\infty} \left(\frac{1}{x+n} + \frac{1}{x-n}\right)$$

$$= \pi \cot(\pi x). \tag{3.6}$$

This is the well-known decomposition of the cotangent by partial fractions. Although the reader should be familiar with this from calculus, we will deduce it from Euler's product representation of the sine because the steps in this deduction are typical and use arguments that will reoccur. Euler's general idea for representing functions as products is to find a representation analogous to the decomposition of a polynomial in linear factors of the form $x - \alpha_\nu$, where the α_ν are the roots of the polynomial. Of course, the product will usually not be finite, and one has to deal with questions of convergence. Once a representation of the product has been found, taking logarithms—which is permitted under certain conditions—transforms the product into a series. For $\sin(\pi x)$, the roots of which are all integers, Euler found the following product representation.

$$\sin \pi x = x \prod_{n \neq 0} \left(1 - \frac{x}{n} \right)$$

$$= x \prod_{n=1}^{\infty} \left(1 - \frac{x^2}{n^2} \right). \tag{3.7}$$

The expression on the right-hand side converges absolutely. This product leads to the representation of the cotangent by partial fractions. $\pi \cot(\pi x)$ is the logarithmic derivative of $\sin(\pi x)$. Thus, taking logarithms (which can be done term by term because the product converges absolutely) the product representation of $\sin \pi x$ becomes

$$\log(\sin \pi x) = \log x + \sum_{n=1}^{\infty} \log\left(1 - \frac{x^2}{n^2} \right),$$

and by term by term differentiation,

$$\pi \cot(\pi x) = \frac{1}{x} + \sum_{n=1}^{\infty} \frac{-2x}{n^2(1 - x^2/n^2)}$$

$$= \frac{1}{x} + \sum_{n=1}^{\infty} \left(\frac{1}{x+n} + \frac{1}{x-n} \right).$$

Let us mention that the theory of periodic and trigonometric functions can be developed from the above definition of the cotangent (cf. André Weil, *Elliptic Functions According to Eisenstein and Kronecker*, Springer-Verlag, Ergebnisse der Mathematik, 1976).

PROOF OF (3.4). Let us substitute $x = 2iz$ in $x/(e^x - 1)$. Then

$$\frac{2iz}{e^{2iz} - 1} = \sum_{k=0}^{\infty} \frac{B_k}{k!} (2iz)^k = \sum_{k=0}^{\infty} \frac{2^k i^k B_k}{k!} z^k$$

$$= 1 - iz + \sum_{k=1}^{\infty} \frac{2^{2k}(-1)^k B_{2k}}{(2k)!} z^{2k},$$

because the B_k vanish for all odd $k > 1$. On the other hand,

$$z \cot z = z \frac{\cos z}{\sin z} = z \frac{(1/2)(e^{iz} + e^{-iz})}{(1/2i)(e^{iz} - e^{-iz})}$$

$$= iz \frac{e^{iz} + e^{-iz}}{e^{iz} - e^{-iz}} = iz \frac{e^{2iz} + 1}{e^{2iz} - 1}$$

$$= iz \frac{2 + e^{2iz} - 1}{e^{2iz} - 1} = \frac{2iz}{e^{2iz} - 1} + iz.$$

From this last equation it follows that

$$z \cot z = 1 + \sum_{k=1}^{\infty} \frac{2^{2k}(-1)^k B_{2k}}{(2k)!} z^{2k}.$$

Using the decomposition of the cotangent in partial fractions (with $z = \pi x$) yields

$$z \cot z = \pi x \cot(\pi x) = 1 + x \sum_{n=1}^{\infty} \left(\frac{1}{x+n} + \frac{1}{x-n} \right)$$

$$= 1 + \frac{z}{\pi} \sum_{n=1}^{\infty} \left(\frac{1}{z/\pi + n} + \frac{1}{z/\pi - n} \right)$$

$$= 1 + z \sum_{n=1}^{\infty} \left(\frac{1}{z + n\pi} + \frac{1}{z - n\pi} \right)$$

$$= 1 + 2 \sum_{n=1}^{\infty} \frac{z^2}{z^2 - n^2\pi^2} = 1 - 2 \sum_{n=1}^{\infty} \frac{z^2}{n^2\pi^2} \left(\frac{1}{1 - z^2/n^2\pi^2} \right)$$

$$= 1 - 2 \sum_{n=1}^{\infty} \frac{z^2}{n^2\pi^2} \sum_{k=0}^{\infty} \frac{z^{2k}}{n^{2k}\pi^{2k}} \qquad \text{(geometrical series)}$$

$$= 1 - 2 \sum_{k=0}^{\infty} \left(\sum_{n=1}^{\infty} \frac{1}{n^{2k+2}} \right) \frac{z^{2k+2}}{\pi^{2k+2}} \qquad \text{(absolute convergence)}$$

$$= 1 - 2 \sum_{k=1}^{\infty} \left(\sum_{n=1}^{\infty} \frac{1}{n^{2k}} \right) \frac{z^{2k}}{\pi^{2k}}.$$

Comparing the coefficients of both representations of $z \cot z$ one obtains

$$\sum_{n=1}^{\infty} \frac{1}{n^{2k}} = \frac{2^{2k-1}\pi^{2k}(-1)^{k+1} B_{2k}}{(2k)!}.$$

Since this series is positive, the B_{2k} have alternating signs, which completes our proof.

(3.8) **Corollary.** $|B_{2k}|$ and even $\sqrt[2k]{|B_{2k}|} \to \infty$ with $k \to \infty$.

PROOF. Let us write $G_k := \sum_{n=1}^{\infty}(1/n^{2k})$. Then,

$$0 < G_k - 1 = \frac{1}{2^{2k}} \sum_{n=2}^{\infty} \left(\frac{2}{n}\right)^{2k} \leqq \frac{1}{2^{2k}} \sum_{n=2}^{\infty} \left(\frac{2}{n}\right)^2 = \frac{G_1 - 1}{2^{2k-2}},$$

so that $\lim_{k \to \infty} G_k = 1$, i.e., by (3.4)

$$\lim_{k \to \infty} \frac{|B_{2k}|}{(2k)!} 2^{2k-1} \pi^{2k} = 1.$$

Then

$$\lim_{k \to \infty} \sqrt[2k]{\frac{(2k)!}{|B_{2k}|}} = 2\pi.$$

This proves our corollary.

Euler's life, as opposed to Fermat's, was very eventful. Leonhard Euler was born in Basel in 1707. His father, Paul Euler, a parson interested in mathematics, who had taken courses from Jacob Bernoulli, was his son's first teacher. In 1720, not quite 14 years old, Euler became a student at the University of Basel, first in theology and philosophy. He also attended mathematical lectures given by Johann I. Bernoulli. Though he passed his examination in philosophy, his main interest was mathematics. At the age of 18, Euler published his first mathematical paper. When two of Johann I. Bernoulli's sons, Daniel and Nicholas II., were called to St. Petersburg by Catherine I to be members of the newly founded Academy, created according to plans of Peter the Great, they tried to find a position for their friend Euler. They were partially successful, since Euler was invited to join the medical department of the Academy in 1726. After quickly studying some physiology, he arrived in Petersburg after a fairly arduous trip (from April 5–May 24, 1727). Contrary to his expectations he was immediately appointed adjunct to the mathematical class, and in 1731, appointed Professor of Natural Science, and in 1733, Professor of Mathematics as the successor of Daniel Bernoulli. Euler devoted much of his time to applied sciences (physics and engineering, maps, navigation, shipbuilding) and the teaching of mathematics. He also wrote several texts in mathematics and physics. Nonetheless, Euler's most important achievements are in pure mathematics; but even they are often computationally oriented. Before formulating general theorems, Euler used to verify special cases through calculations.

In 1740, the political situation in the Russian capital was very confused. At the same time, Frederick the Great, who had just become King of

Jacob Bernoulli

Daniel Bernoulli   Johann Bernoulli
Leonhard Euler   Jacob Steiner

Joseph Louis Lagrange. Published by permission, Germanisches Museum, Munich.

Adrien Marie Legendre

Prussia, made an effort to revive the Berlin Academy of Science that had been founded by his grandfather Frederick I. Euler was invited to work there, accepted, and with his large family reached Berlin on July 25, 1741. There he was faced with a multitude of problems, partly administrative, but also practical ones such as the assignment to plan the construction of a canal, insurance matters, and ballistics. But he did not neglect mathematics and physics. He published much and maintained a lively correspondence with scientists all over Europe. Frederick the Great and Euler were very different, both intellectually and in their personalities. The King, very interested in literature, music and philosophy, surrounded himself with artists, philosophers, and free thinkers. This was a world into which Euler did not fit at all. Educated as a protestant, he was a devout Christian all his life. It should not come as a surprise that leading personalities at the Court made fun of him when he involved himself, with his Christian background, in the current philosophical quarrels about the so-called doctrine of monads. This hostile atmosphere and his failure to become President of the Academy after Maupertuis died might have contributed to the decision to accept Catherine the Great's invitation to return to the Academy in St. Petersburg in 1766. Euler lived there until his death on September 18, 1783. Even though he lost his eyesight in 1771, he continued to be incredibly productive. The publication of his works, started in 1911 by the Schweizerische Naturforschende Gesellschaft, has not yet been completed! His mathematical works alone occupy more than one yard of shelf space. Euler made meaningful contributions in every field of mathematics in which he worked. Doubtlessly, his most important achievements are in analysis (infinite series, theory of functions, differential and integral calculus, differential equations, calculus of variations). We have already mentioned that Euler's applications of infinite series to different number theoretical problems was of principal importance. We will now study this.

The study of the series Σn^{-2k} leads to series of the form Σn^{-s}, $s \in \mathbb{N}$. The case where $s = 2k + 1$ runs into major difficulties. Even today, no explicit formula like (3.4) is known for the corresponding series. Minkowski discovered interesting and very different interpretations for these expressions (see Ch. 10). Euler was probably the first to see that these series can be applied to number theory. His proof of the existence of infinitely many primes uses the divergence of the harmonic series Σn^{-1}: If we assume that there are only finitely many prime numbers, the product

$$\prod_p \left(1 - \frac{1}{p}\right)^{-1},$$

as p runs through the prime numbers, is finite. Expanding every factor in a geometrical series and using the so-called fundamental theorem of arithmetic which says that every natural number can uniquely be written as product of powers of primes, one obtains

$$\infty > \prod_p \left(1 - \frac{1}{p} \right)^{-1} = \prod_p (1 + p^{-1} + p^{-2} + \cdots) = \sum_{n=1}^{\infty} \frac{1}{n} = \infty,$$

a contradiction!

L. P. G. Dirichlet (1805–1859) systematically introduced analytical methods in number theory. Among other things, he investigated the series Σn^{-s} for real s. B. Riemann (1826–1866) allowed complex s.

For real $s \geqslant 1 + \epsilon$ ($\epsilon > 0$), $\Sigma n^{-(1+\epsilon)}$ majorizes Σn^{-s}. Thus Σn^{-s} is uniformly convergent for $s \geqslant 1 + \epsilon$ and is a continuous function of s because of the continuity of its summands. The function represented by this series is called the Zeta-function, and is denoted by $\zeta(s)$. One obtains

$$\frac{1}{s-1} \leq \int_1^{\infty} \frac{1}{x^s}\, dx \leq \zeta(s) \leq 1 + \int_1^{\infty} \frac{1}{x^s}\, dx \leq 1 + \frac{1}{s-1},$$

and consequently

$$\lim_{s \downarrow 1} \zeta(s)(s-1) = 1, \quad \text{specifically} \quad \lim_{s \downarrow 1} \zeta(s) = \infty. \tag{3.9}$$

The Zeta-function has a first order pole at $s = 1$.

If one now expands each factor of the infinite product $\prod_p (1 - p^{-s})^{-1}$, p running over the primes, in a geometrical series and again uses the fundamental theorem of arithmetic, one obtains, analogous to the above, Euler's product representation for the Zeta-function for all real $s > 1$:

$$\zeta(s) = \sum_{n=1}^{\infty} \frac{1}{n^s} = \prod_p (1 - p^{-s})^{-1}. \tag{3.10}$$

(3.11) **Theorem.** $\sum_p 1/p$ is divergent.

PROOF. $\lim_{s \downarrow 1}(\log \zeta(s)) = \infty$ because $\lim_{s \downarrow 1}\zeta(s) = \infty$. Because of

$$\log \zeta(s) = \log\left(\prod_p (1 - p^{-s})^{-1} \right) = \sum_p \log\left(\frac{1}{1 - p^{-s}} \right)$$

$$= \sum_p \sum_{n=1}^{\infty} \frac{p^{-ns}}{n} \quad \text{(logarithmic series)}$$

$$= \sum_p p^{-s} + \sum_p \sum_{n=2}^{\infty} \frac{p^{-ns}}{n},$$

it suffices to show that $\sum_p \sum_{n=2}^{\infty} p^{-ns}/n$ converges. This can be seen from the following rough estimate:

$$\sum_p \sum_{n=2}^{\infty} \frac{1}{np^{ns}} < \sum_p \sum_{n=2}^{\infty} \frac{1}{p^{ns}} = \sum_p \left(\frac{1}{1 - p^{-s}} \right) \frac{1}{p^{2s}} = \sum_p \frac{1}{p^s(p^s - 1)}$$

$$\leq \sum_p \frac{1}{p(p-1)} \leq \sum_{n=2}^{\infty} \frac{1}{n(n-1)} = 1.$$

From (3.11) one obtains a first statement about the distribution of prime numbers: the prime numbers are denser than the squares since

$$\sum_{n=1}^{\infty} \frac{1}{n^2} = \frac{\pi^2}{6} < \infty$$

by (3.4). The series $\sum_p 1/p$ diverges very slowly. The partial sum after the first 50 million terms is still less than 4.

Another typical example for Euler's way of thinking is the following attempt to prove the four-square theorem. The problem fascinated him over several decades but he never found a complete proof; however, he improved and simplified the first proof of the theorem, due to Lagrange. Euler's approach tries to use the function

$$f(x) = 1 + x + x^4 + x^9 + x^{25} + \cdots$$

which is defined for $|x| < 1$. After expanding the function $f(x)^4$ in a power series

$$f(x)^4 = \sum_{n=0}^{\infty} \tau(n) x^n$$

and comparing coefficients, $\tau(n)$ will denote the number of ways n can be written as a sum of four squares. To prove Fermat's statement, one only has to show that $\tau(n) > 0$. It is quite difficult to do this; only many years later did C. G. J. Jacobi succeed, using the theory of elliptic functions. We do, however, see how Euler nearly magically transforms a purely arithmetical question into an analytical problem. In fact, Euler's idea is more general, as we will now see in our discussion of a similar problem. A *partition* of a natural number is a representation as a sum of natural numbers. Two partitions are the same if they differ only in the sequence of their summands: consequently, we can always assume in a partition $n = n_1 + \cdots + n_k$ that $n_1 \geqslant n_2 \geqslant \cdots \geqslant n_k$. Let $p(n)$ be the number of partitions of n. For example, $p(2) = 2$, $p(3) = 3$, $p(4) = 5$, $p(5) = 7$. In 1663, Leibniz suggested investigating these partitions in a letter to Johann I. Bernoulli. It is extremely difficult to compute $p(n)$ for arbitrary n. Like $\tau(n)$ considered above, $p(n)$ is an arithmetical function, i.e., a function $f : \mathbb{N} \rightarrow \mathbb{N}$. Euler assigns to every such function a series

$$F(x) := \sum_{n=0}^{\infty} f(n) x^n, \qquad f(0) := 1 \qquad (3.12)$$

F is called the *generating function* of f. If $f(n)$ does not approach infinity too rapidly with n, this series has a positive radius of convergence. When f is the partition function, this series converges for $|x| < 1$. Euler shows:

(3.13) **Theorem.** *For* $|x| < 1$,

$$\sum_{n=0}^{\infty} p(n)x^n = \prod_{m=1}^{\infty} \frac{1}{1-x^m}$$

(*with* $p(0) = 1$).

PROOF. First, we expand every factor in a geometrical series and obtain

$$\prod_{m=1}^{\infty} \frac{1}{1-x^m}$$

$$= (1 + x + x^2 + \cdots)(1 + x^2 + x^4 + \cdots)(1 + x^3 + x^6 + \cdots)\cdots.$$

Disregarding questions of convergence, we multiply the series on the right-hand side as if they were polynomials and order them according to powers of x. Then we obtain a power series of the form

$$\sum_{k=0}^{\infty} a(k)x^k \qquad (a(0) := 1).$$

Now we have to show that $a(k) = p(k)$. Let x^{k_1} be a term of the first series, x^{2k_2} a term of the second, and, generally, x^{mk_m} a term of the mth series. The product of these terms is

$$x^{k_1}x^{2k_2}\dots x^{mk_m} = x^k$$

with

$$k = k_1 + 2k_2 + \cdots + mk_m.$$

This last expression is a partition of k. Any term gives us a partition of k; conversely, any partition of k corresponds to a term. This relation is one to one; so $a(k) = p(k)$. This is not yet a complete proof. We will now fill in the gaps. First let $x \in [0, 1)$. We introduce the functions

$$G_m(x) = \prod_{k=1}^{m} \frac{1}{1-x^k}, \qquad G(x) = \prod_{k=1}^{\infty} \frac{1}{1-x^k} = \lim_{m\to\infty} G_m(x).$$

The product defining G converges for $x \in [0, 1)$ because the series $\sum x^k$ do. For fixed x in $[0, 1)$, the series $G_m(x)$ grows monotonically. Therefore, $G_m(x) \leqslant G(x)$ for fixed $x \in [0, 1)$ and every m. Since $G_m(x)$ is a product of a finite number of absolutely convergent series, $G_m(x)$ is absolutely convergent and can be written as

$$G_m(x) = \sum_{k=0}^{\infty} p_m(k)x^k,$$

where $p_m(k)$ denotes the number of partitions of k into parts not greater than m ($p_m(0) := 1$). For $m \geqslant k$, $p_m(k) = p(k)$. Since $p_m(k) \leqslant p(k)$ one obtains $\lim_{m\to\infty} p_m(k) = p(k)$.

Let us now decompose $G_m(x)$ into two parts:

$$G_m(x) = \sum_{k=0}^{m} p_m(k)x^k + \sum_{k=m+1}^{\infty} p_m(k)x^k$$

$$= \sum_{k=0}^{m} p(k)x^k + \sum_{k=m+1}^{\infty} p_m(k)x^k.$$

Because $x \geqslant 0$,

$$\sum_{k=0}^{m} p(k)x^k \leq G_m(x) \leq G(x).$$

Consequently $\sum_{k=0}^{\infty} p(k)x^k$ converges; because $p_m(k) \leqslant p(k)$,

$$\sum_{k=0}^{\infty} p_m(k)x^k \leq \sum_{k=0}^{\infty} p(k)x^k \leq G(x).$$

Consequently, the series $\sum_{k=0}^{\infty} p_m(k)x^k$ converges uniformly for all m and

$$G(x) = \lim_{m \to \infty} G_m(x) = \lim_{m \to \infty} \sum_{k=0}^{\infty} p_m(k)x^k$$

$$= \sum_{k=0}^{\infty} \lim_{m \to \infty} p_m(k)x^k$$

$$= \sum_{k=0}^{\infty} p(k)x^k.$$

This proves Euler's formula for $x \in [0, 1)$. By analytic continuation, the proof follows for $x \in (-1, 1)$.

Let $q(n)$ be the number of partitions of n in odd summands and $r(n)$ the number of partitions of n into different summands. Then the generating functions of q and r can be found in a similar way.

(3.14) **Theorem** (Euler).

$$\frac{1}{(1-x)(1-x^3)(1-x^5) \ldots}$$

is a generating function of q and $(1 + x)(1 + x^2)(1 + x^3) \ldots$ is a generating function of r.

The first statement can be proved in a way similar to (2.4); the second is trivial.

(3.15) **Theorem** (Euler). $q(n) = r(n)$.

The proof is easy with the help of the corresponding generating functions. We just have to show that they coincide. The statement follows by comparing coefficients. In fact,

$$(1 + x)(1 + x^2)(1 + x^3) \cdots = \frac{(1 - x^2)(1 - x^4)(1 - x^6)}{(1 - x)(1 - x^2)(1 - x^3)} \cdots$$

$$= \frac{1}{(1 - x)(1 - x^3)(1 - x^5)} \cdots .$$

Without generating functions the proof of the theorem is not obvious; cf. Hardy and Wright, *The Theory of Numbers*.

We have just seen a fine example of the power of this method, and we end with another theorem of Euler which, however, we will not prove.

Let us look at $\prod_{m=1}^{\infty}(1 - x^m)$, the reciprocal of the generating function of p. The first terms are

$$(1 - x)(1 - x^2)(1 - x^3)(1 - x^4)(1 - x^5)(1 - x^6)(1 - x^7) \cdots$$

$$= 1 - x - x^2 + x^5 + x^7 - x^{12} - x^{15} + - \cdots .$$

The series does not follow an obvious law, and Euler certainly calculated a great number of terms before he found it. A few years later he proved:

(3.16) **Theorem.**

$$\prod_{m=1}^{\infty} (1 - x^m) = \sum_{k=-\infty}^{+\infty} (-1)^k x^{(3k^2+k)/2}$$

$$:= \sum_{k=0}^{\infty} (-1)^k \left(x^{(3k^2-k)/2} + x^{(3k^2+k)/2} \right).$$

It was Jacobi who gave the first "natural" proof of this result, again within the framework of the theory of elliptic functions. There is an attractive combinatorial proof of F. Franklin (1881) to be found in Hardy and Wright.

Obviously, what has been discussed so far in this chapter can be characterized by the use of analytical methods and might be said to belong to analysis rather than to number theory. Number theory, in a way, did not exist when Euler began his work, since Fermat had not left any proofs. Initially, Euler was quite isolated; only later did Lagrange join him as a versatile and knowledgeable partner. It is difficult to realize today what kind of obstacles Euler faced, obstacles which we can overcome easily today with the help of simple algebraic concepts such as the theory of groups. André Weil, one of the most eminent mathematicians of our time made the following comments on Euler's number-theoretical work: "One must realize that Euler had absolutely nothing to start from except Fermat's mysterious-looking statements ... Euler had to reconstruct everything from scratch ... ". However, one would not do justice to Euler's versatility if one tried to pin him down too tightly, be it with regard to the problems he treated or to the methods he used. On the contrary, it is significant and typical for Euler that he was interested in everything and pursued so many different and disparate questions with the enthusiasm of the natural philosopher of his period.

To conclude this chapter, we will mention a few more results which show the breadth of Euler's mathematical work. Some reflect deep insights, others are just curiosities. But everything has some connection to number theory and we will discuss some of these results later in more detail.

There are formulas, easy to understand but not so simple to prove, such as

$$\int_0^\infty \frac{\sin x}{x}\, dx = \frac{\pi}{2}\,,$$

$$\int_0^\infty \sin x^2\, dx = \frac{1}{2}\sqrt{\frac{\pi}{2}}\,,$$

$$1 - \frac{1}{2} + \frac{1}{4} - \frac{1}{5} + \frac{1}{7} - \cdots = \frac{\pi}{3\sqrt{3}}\,.$$

Then there is another formula which is very difficult to prove and seems very obscure at first sight:

$$\frac{1 - 2^{m-1} + 3^{m-1} - 4^{m-1} + \cdots}{1 - 2^{-m} + 3^{-m} - 4^{-m} + \cdots} = \frac{1 \cdot 2 \cdot 3 \cdots (m-1)(2^m - 1)}{(2^{m-1}1)\pi^m} \cos \frac{m\pi}{2}\,.$$

(*Remarques sur un beau rapport entre les series des Puissances tant directes que reciproques*, Opera (1), Vol. 15, p. 83). If one looks at it more closely, one sees that this is the functional equation for the Zeta-function. Then, in his correspondence and in his papers, there are various strictly arithmetical theorems for which Euler does not have a proof and which he does not even state precisely, among them the following. The numbers $d = 1, 2, 3, 4,$ $5, 6, 7, 8, 9, 10, 12, 13, 15, 16, 18, 21, \ldots, 1320, 1365, 1848$ (altogether 65) have the following property: If $ab = d$ and if a number can be uniquely written in the form $ax^2 + by^2$ with ax, by relatively prime, then this number is of the form p, $2p$, or 2^k, where p is a prime number. Specifically, any odd number > 1 that can be written uniquely in this way is prime; Euler calls these numbers *numeri idonei* because they can be used for primality tests. He gives the following application for $d = 57$. 1000003 is a prime number because it can be written uniquely as

$$19 \cdot 8^2 + 3 \cdot 577^2.$$

$d = 1848$ yields the prime number 18518809 with the unique representation

$$197^2 + 1848 \cdot 100^2.$$

It is still unsolved whether Euler's 65 numbers are the only *numeri idonei*. (Only for the cases $d = 1, 2, 3$ did Euler show that they have the required property. In the case of $d = 1$ there is an obvious connection to Fermat's theorem (2.3).)

The following is a decidedly curious statement. $x^2 + x + 41$ is a prime number for $x = 0, 1, 2, \ldots, 39$. Of course, one can check this easily, but how does one find such a result and what is the real reason? (The field $\mathbb{Q}(\sqrt{-163})$ has class number 1.)

It is equally easy to check the following purely algebraic formula (with xx in Euler's notation, instead of x^2):

$$(aa + bb + cc + dd)(pp + qq + rr + ss) = xx + yy + zz + vv$$

with

$$x = ap + bq + cr + ds,$$
$$y = aq - bp \pm cs \mp dr,$$
$$z = ar \mp bs - cp \pm dq,$$
$$v = as \pm br \mp cq - dp.$$

Obviously, this means that the product of two sums of four squares is the sum of four squares. This means that one can confine oneself to primes in the proof of Fermat's four-square theorem.

We conclude this list with the statement of the law of quadratic reciprocity which Euler found but could not prove: An odd prime number s is a square modulo an odd prime number p if and only if $(-1)^{1/2(p-1)}p$ is a square modulo s.

References

L. Euler: Introductio in Analysin infinitorum, *Opera Omnia* (1), Vol. 8.
Hardy and Wright, specifically, Chaps. 17 and 19.
A. Weil: Two lectures on Number Theory, Past and Present.
L. Kronecker: *Zur Geschichte des Reziprozitätsgesetzes*, Werke II, 1–10.
J. Steinig: On Euler's idoneal numbers, Elem. der Math. 21 (1966), 73–88.
Th. L. Heath: see references to Chap. 2.
J. E. Hofmann: see references to Chap. 2.
A. P. Youschkevitch: Euler, Leonhard (in: *Dictionary of Scientific Biography*).
N. Fuss: Lobrede auf Herrn Leonhard Euler, in: Euler, *Opera Omnia* (1), Vol. 1.
Euler–Goldbach: *Briefwechsel* (Correspondence).

CHAPTER 4
Lagrange

Joseph Louis Lagrange lived from 1736 to 1813. Born in Turino, he had both French and Italian ancestors. His family was well off but Lagrange's father lost the family fortune in risky financial transactions. This is said to have prompted Lagrange to remark, "Had I inherited a fortune I would probably not have fallen prey to mathematics." (cf. E. T. Bell, *Men of Mathematics*). As a youth Lagrange was more interested in classical languages than in mathematics, but his interest in mathematics was stirred by a paper by Halley, the friend of Newton. In a short time he acquired a deep knowledge of analysis; only 19 years old, he became Professor at the Royal School of Artillery in Turino. Lagrange stayed there for about 10 years. His reputation as a mathematician grew quickly, mainly by basic contributions to analysis, specifically the calculus of variations, the theory of differential equations, and mechanics. This combination of mathematics and mechanics or, more generally, theoretical physics, is typical of the eighteenth century. Mathematics was not viewed as an end in itself but mostly as a tool for understanding nature. In 1766, d'Alembert was instrumental in bringing Lagrange to succeed Euler at the Berlin Academy of Science. Financial conditions in Berlin were very good; moreover, he could devote himself exclusively to his mathematical work. Lagrange stayed there until 1787 when he moved to the Academie Française in Paris. At that time, soon after Euler's death, he was recognized as the most important living mathematician. Though Lagrange had had close ties to the French royal family he was not persecuted during the French Revolution. Altogether, the sciences gained importance during the era of the French Revolution and Napoleon. Lagrange's authority transcended the sphere of science. He was a Senator of the Empire and in fact received a state burial in the Pantheon.

While he was in Paris, Lagrange was quite involved in the problems of teaching mathematics and other areas of science. For a time, he even seems to have dropped his interest in mathematics. Lagrange's number-theoretical papers belong to the Berlin era, mainly to the years 1766–1777. Lagrange's main inspiration seems to have been Euler's work which he read very carefully. Though there is an extensive correspondence between Euler and Lagrange, they never met.

Euler was not really successful in treating Fermat's problems. In spite of great efforts he gave a complete proof, after several unsuccessful attempts, only of the two-square theorem. Euler's contributions to the four-square theorem or to the theory of the equations $x^3 = y^2 + 2$ or $x^3 + y^3 = z^3$ were almost successful, but serious gaps remained. Euler's real achievement was the presentation of many examples and the use of analytical methods.

Lagrange is Fermat's true successor in number theory. He was the first to give proofs for a series of Fermat's propositions and did so without leaving the realm of arithmetic; many of these techniques were his own. Three of Lagrange's (not very numerous) papers in number theory are particularly important:

"Solution d'un problème d'arithmétique" (1768, Oeuvres de Lagrange I, 671–731). Lagrange treats the equation $x^2 - dy^2 = 1$ (see (2.5)).

"Démonstration d'un théorème d'arithmétique" (1770, Oeuvres III, 189–201). This paper contains the first proof of the four-square theorem (2.4).

"Recherches d'arithmétique" (1773, Oeuvres III, 695–795). Lagrange develops the theory of binary quadratic forms and derives from the general theory, among other things, Fermat's theorems about the representation of prime numbers by $x^2 + 2y^2$ and $x^2 + 3y^2$.

We are particularly interested in this third paper because it is the first work to develop systematically and coherently a complete arithmetical theory, going much further than the individual problems which are discussed by Fermat and Euler. The importance of this step cannot be overestimated for the further development of number theory and algebra. About 25 years later, Gauss considerably expanded the theory of binary quadratic forms. We will discuss this below, though we will use some of Gauss's terminology in this chapter.

This might be a good occasion to mention that it is often difficult or sometimes nearly impossible to credit a mathematical result to just one mathematician. Often A discovers a theorem, B gives a partial proof, C proves it completely, and D generalizes it. This gives us a certain freedom in deciding in what context to discuss a specific theorem—and this freedom we will often exploit.

Returning to Lagrange, we note that his papers are written in "contemporary" mathematical style. They are very readable, even exemplary in their clear and well-organized presentation. We will now give a systematic development of the foundations of the theory of binary quadratic forms,

following Lagrange's exposition. Let us start with a free translation of
excerpts from the introduction of the paper "Recherches d'arithmétique":

> These investigations are concerned with the numbers that can be written in
> the form
> $$Bt^2 + Ctu + Du^2$$
> where B, C, D are integers and t, u are also integers but variable. Thus, I will
> determine those forms which represent numbers whose divisors can be
> represented in the same way; later I will give a technique which permits us to
> reduce these forms to their smallest number. This will lead to a table for
> practical use; I will show how to use this table in the investigation of the
> divisors of a number. I will finally give proofs for several theorems about
> prime numbers of the form $Bt^2 + Ctu + Du^2$; some of these theorems are
> known but without proof and others are completely new.

Thus the author studies the quadratic forms

$$q(x, y) = ax^2 + bxy + cy^2.$$

Certain quadratic forms, namely $x^2 + y^2, x^2 + 2y^2, x^2 + 3y^2, x^2 - dy^2$, were
already treated by Fermat (see Chapter 2). First, Lagrange investigates the
divisors of a number represented by $ax^2 + bxy + cy^2$. One says that a
number m can be *represented* by this form if the equation

$$m = ax^2 + bxy + cy^2$$

is solvable in integers. Lagrange proves the following theorem; the proof is
taken nearly word for word from his paper.

(4.1) Theorem. *Let r be a divisor of a number that can be represented by the
form $ax^2 + bxy + cy^2$ with $x = x_0$, $y = y_0$ relatively prime. Then r can be
represented by a form $AX^2 + BXY + CY^2$ with $X = X_0$, $Y = Y_0$ relatively
prime, and $4AC - B^2 = 4ac - b^2$.*

PROOF. Let

$$rs = ax^2 + bxy + cy^2.$$

Let t be the greatest common divisor of s and y, i.e., $s = tu$, $y = tX$ with u
and X relatively prime. This leads to

$$rtu = ax^2 + btxX + ct^2X^2$$

which means that t divides ax^2. By our assumptions, x and y are relatively
prime, consequently, x and t are as well. This means that t divides a, i.e.,
$a = et$. Dividing by t gives

$$ru = ex^2 + bxX + ctX^2.$$

Since u and X are relatively prime, we can write x in the form

$$x = uY + wX.$$

Substituting this in the last equation, one obtains

$$ru = e(uY + wX)^2 + b(uY + wX)X + ctX^2$$
$$= (ew^2 + bw + ct)X^2 + (2euw + bu)XY + eu^2Y^2.$$

The first summand must be divisible by u. Since u and X are relatively prime, u is a divisor of $ew^2 + bw + ct$. Setting

$$A := \frac{ew^2 + bw + ct}{u}, \qquad B := 2ew + b, \qquad C := eu,$$

one obtains, as required,

$$r = AX^2 + BXY + CY^2.$$

X and Y are relatively prime, and it is easy to check that

$$4AC - B^2 = 4ac - b^2.$$

We say that a number m is *properly* represented by a (binary) quadratic form q if the equation $m = q(x, y)$ can be solved in relatively prime integers. m is called a divisor of q if m is a divisor of a number that can be properly represented by q. The expression $4ac - b^2$ is called the discriminant* of the form $ax^2 + bxy + cy^2$. Now we can reformulate theorem (4.1):

(4.1)′ **Theorem.** *If m is a divisor of a quadratic form, m can be properly represented by a quadratic form with the same discriminant.*

In what follows we consider, instead of $ax^2 + bxy + cy^2$, the more special quadratic form

$$ax^2 + 2bxy + cy^2.$$

In this, we follow Lagrange and, later, Gauss. (4.1) and (4.1)′ are valid for this new form, too, because if b is even B also is even as $B = 2ew + b$. Using matrices, one can write

$$ax^2 + 2bxy + cy^2 = (x, y)\begin{pmatrix} a & b \\ b & c \end{pmatrix}\begin{pmatrix} x \\ y \end{pmatrix}.$$

One sees that the form can completely be described by a 2×2 matrix with integral entries a, b, c. Depending on the context, we will identify the form with the matrix. Frequently, we will use the abbreviations $q(x, y) = ax^2 + 2bxy + cy^2$ and

$$\Delta := \det\begin{pmatrix} a & b \\ b & c \end{pmatrix} = ac - b^2.$$

Throughout, we will assume $\Delta \neq 0$.

Two forms

$$ax^2 + 2bxy + cy^2, \qquad AX^2 + 2BXY + CY^2$$

are equivalent (or isomorphic) if they can be transformed into each other with an invertible integral linear substitution of variables, i.e., if

$$\begin{aligned} X &= \alpha x + \beta y \\ Y &= \gamma x + \delta y \end{aligned} \quad \text{with} \quad \begin{pmatrix} \alpha & \beta \\ \gamma & \delta \end{pmatrix} \in GL(2, \mathbb{Z}).$$

---

*(Translator's note). This is occasionally called the *determinant*, but for quadratic forms discriminant is more common.

The forms are properly equivalent if $\left(\begin{smallmatrix} \alpha & \beta \\ \gamma & \delta \end{smallmatrix}\right) \in SL(2, \mathbb{Z})$. Here $GL(2, \mathbb{Z})$ and $SL(2, \mathbb{Z})$ denote, as usual, the groups of invertible integral 2×2 matrices and the group of invertible integral 2×2 matrices with determinant 1.

Using matrix notation, one has

$$\begin{pmatrix} X \\ Y \end{pmatrix} = \begin{pmatrix} \alpha & \beta \\ \gamma & \delta \end{pmatrix} \begin{pmatrix} x \\ y \end{pmatrix}.$$

This leads to

$$(X, Y)\begin{pmatrix} A & B \\ B & C \end{pmatrix}\begin{pmatrix} X \\ Y \end{pmatrix} = (x, y)\begin{pmatrix} \alpha & \gamma \\ \beta & \delta \end{pmatrix}\begin{pmatrix} A & B \\ B & C \end{pmatrix}\begin{pmatrix} \alpha & \beta \\ \gamma & \delta \end{pmatrix}\begin{pmatrix} x \\ y \end{pmatrix}$$

$$= (x, y)\begin{pmatrix} a & b \\ b & c \end{pmatrix}\begin{pmatrix} x \\ y \end{pmatrix}.$$

Thus, two matrices $\left(\begin{smallmatrix} a & b \\ b & c \end{smallmatrix}\right), \left(\begin{smallmatrix} A & B \\ B & C \end{smallmatrix}\right)$ define equivalent (or properly equivalent) forms if and only if there is $T \in GL(2, \mathbb{Z})$ ($T \in SL(2, \mathbb{Z})$) such that

$$\begin{pmatrix} A & B \\ B & C \end{pmatrix} = T'\begin{pmatrix} a & b \\ b & c \end{pmatrix}T,$$

where T' is the transpose of T. "Equivalence" and "proper equivalence" determine equivalence relations. Equivalent forms represent the same numbers and have the same discriminant because of $(\det T)^2 = 1$. We call a form $q(x, y) = ax^2 + 2bxy + cy^2$ positive (negative) if $q(x, y) \geqslant 0$ ($\leqslant 0$) for all $x, y \in \mathbb{Z}$. If a form is either positive or negative it is called definite; otherwise it is indefinite.

It is easy to see that a form q is positive if and only if Δ is positive and $a > 0$; q is negative if and only if Δ is positive and $a < 0$. It is indefinite if and only if $\Delta < 0$. To prove this, write

$$ax^2 + 2bxy + cy^2 = a\left(x + \frac{b}{a}y\right)^2 + cy^2 - \frac{b^2}{a}y^2$$

and test the three cases.

Below, we consider only definite and, without loss of generality, positive forms.

Lagrange's next result is the following; it is of fundamental importance for the whole theory.

(4.2) **Theorem.** *A positive form q is properly equivalent to a so-called reduced form, i.e., a form that can be described by a matrix $\left(\begin{smallmatrix} a & b \\ b & c \end{smallmatrix}\right)$ with the entries satisfying*

$$-\frac{a}{2} < b \leqq \frac{a}{2}, \qquad a \leqq c, \quad and \quad 0 \leqq b \leqq \frac{a}{2} \qquad if \quad a = c.$$

These conditions uniquely determine the matrix. Moreover,

$$a \leqslant 2\sqrt{\frac{\Delta}{3}},$$

where Δ is the discriminant of q.

PROOF. Suppose q is described by the matrix $\left(\begin{smallmatrix} A & B \\ B & C \end{smallmatrix}\right)$. Let a be the smallest number which can be represented by q. Thus a can be written

$$a = AX_0^2 + 2BX_0Y_0 + CY_0^2$$

for suitable $X_0, Y_0 \in \mathbb{Z}$. X_0 and Y_0 must be relatively prime. This means that there are $\alpha, \beta \in \mathbb{Z}$ with

$$\alpha X_0 + \beta Y_0 = 1.$$

Then

$$\begin{pmatrix} X_0 & Y_0 \\ -\beta & \alpha \end{pmatrix} \in SL(2, \mathbb{Z})$$

and

$$\begin{pmatrix} X_0 & Y_0 \\ -\beta & \alpha \end{pmatrix}\begin{pmatrix} A & B \\ B & C \end{pmatrix}\begin{pmatrix} X_0 & -\beta \\ Y_0 & \alpha \end{pmatrix} = \begin{pmatrix} a & B' \\ B' & C' \end{pmatrix}$$

with $B', C' \in \mathbb{Z}$. For arbitrary $k \in \mathbb{Z}$, we use the transformation $\left(\begin{smallmatrix} 1 & 0 \\ k & 1 \end{smallmatrix}\right)$ $\in SL(2, \mathbb{Z})$ to obtain

$$\begin{pmatrix} 1 & 0 \\ k & 1 \end{pmatrix}\begin{pmatrix} a & B' \\ B' & C' \end{pmatrix}\begin{pmatrix} 1 & k \\ 0 & 1 \end{pmatrix} = \begin{pmatrix} a & B' + ka \\ B' + ka & * \end{pmatrix}.$$

Let us now pick $k \in \mathbb{Z}$ such that $-a/2 < B' + ka \leqslant a/2$. Setting $b :=$ $B' + ka$, $c := *$, the matrix

$$\begin{pmatrix} a & b \\ b & c \end{pmatrix}$$

is, by construction, properly equivalent to $\left(\begin{smallmatrix} A & B \\ B & C \end{smallmatrix}\right)$ and satisfies the conditions $-a/2 < b \leqslant a/2$ and $a \leqslant c$. Then $a \leqslant c$ since c can be represented (with $x = 0, y = \pm 1$) and a was assumed to be the smallest number that can be represented.

If b is negative for the case $a = c$, we use the matrix $\left(\begin{smallmatrix} 0 & -1 \\ 1 & 0 \end{smallmatrix}\right) \in SL(2, \mathbb{Z})$:

$$\begin{pmatrix} 0 & -1 \\ 1 & 0 \end{pmatrix}\begin{pmatrix} a & b \\ b & a \end{pmatrix}\begin{pmatrix} 0 & 1 \\ -1 & 0 \end{pmatrix} = \begin{pmatrix} a & -b \\ -b & a \end{pmatrix}.$$

Then $-b > 0$.

Now we have to show uniqueness. First we show that if $\left(\begin{smallmatrix} a & b \\ b & c \end{smallmatrix}\right)$ is a reduced form, then a is necessarily the smallest number represented by this form. Hence a is determined uniquely.

Let us now prove this last statement. If $\left(\begin{smallmatrix} a & b \\ b & c \end{smallmatrix}\right)$ is reduced, the form $ax^2 + 2bxy + cy^2$ assumes only values $\geqslant ax^2 \geqslant a$ for $0 < |x| \leqslant |y|$ because of $2bxy + cy^2 \geqslant 0$. If $0 < |y| \leqslant |x|$, then $ax^2 + 2bxy \geqslant 0$, and the form assumes only values $\geqslant cy^2 \geqslant a$. If $x = 0$ or $y = 0$, then again $ax^2 + 2bxy + cy^2 \geqslant a$. One obtains this minimum when $x = \pm 1, y = 0$.

If $a < c$, these are the only values that give the minimum because $|x| > 1, y = 0$ will not give all possible a, and $x, y \neq 0$ leads to

$$ax^2 + 2bxy + cy^2 \geqslant cy^2 > a \qquad \text{for } x \geqslant y \geqslant 1,$$
$$ax^2 + 2bxy + cy^2 > ax^2 \geqslant a \qquad \text{for } 1 \leqslant |x| \leqslant |y|.$$

Consequently, if $\begin{pmatrix} a & B \\ B & C \end{pmatrix}$ is properly equivalent to $\begin{pmatrix} a & b \\ b & c \end{pmatrix}$ in reduced form, say

$$\begin{pmatrix} a & B \\ B & C \end{pmatrix} = \begin{pmatrix} \alpha & \gamma \\ \beta & \delta \end{pmatrix} \begin{pmatrix} a & b \\ b & c \end{pmatrix} \begin{pmatrix} \alpha & \beta \\ \gamma & \delta \end{pmatrix}$$

$$= \begin{pmatrix} a\alpha^2 + 2b\alpha\gamma + c\gamma^2 & * \\ * & * \end{pmatrix},$$

we have $a = a\alpha^2 + 2b\alpha\gamma + c\gamma^2$, i.e., $\gamma = 0$, $\alpha = \pm 1$. The transformation is then given by

$$\begin{pmatrix} a & B \\ B & C \end{pmatrix} = \begin{pmatrix} \pm 1 & 0 \\ \beta & \pm 1 \end{pmatrix} \begin{pmatrix} a & b \\ b & c \end{pmatrix} \begin{pmatrix} \pm 1 & \beta \\ 0 & \pm 1 \end{pmatrix}$$

$$= \begin{pmatrix} a & b \pm \beta a \\ * & * \end{pmatrix}.$$

This leads to $\beta = 0$ because $-a/2 < b$, $B \leqslant a/2$. Also, $B = b$ and consequently $\begin{pmatrix} a & B \\ B & C \end{pmatrix} = \begin{pmatrix} a & b \\ b & c \end{pmatrix}$.

If $a = c$, $0 \leqslant b < a/2$, the smallest a is attained when $x = \pm 1$, $y = 0$ and $x = 0$, $y = \pm 1$. Hence, if $\begin{pmatrix} a & B \\ B & C \end{pmatrix}$ is properly equivalent to $\begin{pmatrix} a & b \\ b & c \end{pmatrix}$ in its reduced form, this equivalence can be expressed by

$$\begin{pmatrix} a & B \\ B & C \end{pmatrix} = \begin{pmatrix} \pm 1 & 0 \\ \beta & \pm 1 \end{pmatrix} \begin{pmatrix} a & b \\ b & a \end{pmatrix} \begin{pmatrix} \pm 1 & \beta \\ 0 & \pm 1 \end{pmatrix} \qquad (*)$$

or

$$\begin{pmatrix} a & B \\ B & C \end{pmatrix} = \begin{pmatrix} 0 & \pm 1 \\ \mp 1 & \beta \end{pmatrix} \begin{pmatrix} a & b \\ b & a \end{pmatrix} \begin{pmatrix} 0 & \mp 1 \\ \pm 1 & \beta \end{pmatrix}. \qquad (**)$$

Then it follows that $B = \pm a\beta + b$ and $B = \pm a\beta - b$. Because of $0 \leqslant b$, $B \leqslant a/2$, it follows that $\beta = 0$, hence $\begin{pmatrix} a & B \\ B & C \end{pmatrix} = \begin{pmatrix} a & b \\ b & c \end{pmatrix}$. If $a = c = 2b$, one has

$$\begin{pmatrix} a & b \\ b & c \end{pmatrix} = b \begin{pmatrix} 2 & 1 \\ 1 & 2 \end{pmatrix},$$

and it suffices to consider the matrix $\begin{pmatrix} 2 & 1 \\ 1 & 2 \end{pmatrix}$. One obtains 2, the minimum of $2x^2 + 2xy + 2y^2$, when $x = \pm 1$, $y = 0$ or $x = 0$, $y = \pm 1$ or $x = \pm 1$, $y = \mp 1$. It follows that $B = b$, $C = c$. The last inequality for a follows from the previous inequalities.

(4.3) **Corollary.** *There are only finitely many proper equivalence classes of positive binary quadratic forms with a given discriminant* Δ.

PROOF. Every proper equivalence class contains the reduced form $\begin{pmatrix} a & b \\ b & c \end{pmatrix}$ with

$$a \leqq 2\sqrt{\frac{\Delta}{3}}, \qquad |b| \leqq \sqrt{\frac{\Delta}{3}}.$$

This means that there are only finitely many possibilities for a and b, hence also for c. We can now devise a table of positive reduced forms (Table 1) (cf., Lagrange, *Oeuvres* III, page 757 or Gauss, Disquisitiones Arithmeticae, art 176).

Table 1

Δ	positive reduced forms			
1	$\begin{pmatrix} 1 & 0 \\ 0 & 1 \end{pmatrix}$			
2	$\begin{pmatrix} 1 & 0 \\ 0 & 2 \end{pmatrix}$			
3	$\begin{pmatrix} 1 & 0 \\ 0 & 3 \end{pmatrix}$	$\begin{pmatrix} 2 & 1 \\ 1 & 2 \end{pmatrix}$		
4	$\begin{pmatrix} 1 & 0 \\ 0 & 4 \end{pmatrix}$	$\begin{pmatrix} 2 & 0 \\ 0 & 2 \end{pmatrix}$		
5	$\begin{pmatrix} 1 & 0 \\ 0 & 5 \end{pmatrix}$	$\begin{pmatrix} 2 & 1 \\ 1 & 3 \end{pmatrix}$		
6	$\begin{pmatrix} 1 & 0 \\ 0 & 6 \end{pmatrix}$	$\begin{pmatrix} 2 & 0 \\ 0 & 3 \end{pmatrix}$		
7	$\begin{pmatrix} 1 & 0 \\ 0 & 7 \end{pmatrix}$	$\begin{pmatrix} 2 & 1 \\ 1 & 4 \end{pmatrix}$		
8	$\begin{pmatrix} 1 & 0 \\ 0 & 8 \end{pmatrix}$	$\begin{pmatrix} 2 & 0 \\ 0 & 4 \end{pmatrix}$	$\begin{pmatrix} 3 & 1 \\ 1 & 3 \end{pmatrix}$	
9	$\begin{pmatrix} 1 & 0 \\ 0 & 9 \end{pmatrix}$	$\begin{pmatrix} 2 & 1 \\ 1 & 5 \end{pmatrix}$	$\begin{pmatrix} 3 & 0 \\ 0 & 3 \end{pmatrix}$	
10	$\begin{pmatrix} 1 & 0 \\ 0 & 10 \end{pmatrix}$	$\begin{pmatrix} 2 & 0 \\ 0 & 5 \end{pmatrix}$		
11	$\begin{pmatrix} 1 & 0 \\ 0 & 11 \end{pmatrix}$	$\begin{pmatrix} 2 & 1 \\ 1 & 6 \end{pmatrix}$	$\begin{pmatrix} 3 & 1 \\ 1 & 4 \end{pmatrix}$	$\begin{pmatrix} 3 & -1 \\ -1 & 4 \end{pmatrix}$
12	$\begin{pmatrix} 1 & 0 \\ 0 & 12 \end{pmatrix}$	$\begin{pmatrix} 2 & 0 \\ 0 & 6 \end{pmatrix}$	$\begin{pmatrix} 3 & 0 \\ 0 & 4 \end{pmatrix}$	$\begin{pmatrix} 4 & 2 \\ 2 & 4 \end{pmatrix}$

We have now enough theoretical results to supplement one of the concrete statements in Chapter 2 (cf. (2.3)).

(4.4) **Theorem** (Fermat). *A natural number $a = b^2c$, c square free, is the sum of two squares if and only if c contains only prime factors of the form $4n + 1$ or 2.*

PROOF. If c is square free and of the given form, it is, by (2.3) and its corollary, the sum of two squares. Consequently a is the sum of two squares. Conversely, let a be the sum of two squares, $a = x_0^2 + y_0^2$. Without loss of generality we can assume that x_0^2 and y_0^2 are relatively prime. Let p be a prime factor of c. Then p is a factor of $x^2 + y^2$; so by (4.1) p can be represented by a positive form with discriminant 1. Up to equivalence there is, however, only one form with discriminant 1. This means that p can be properly represented by $x^2 + y^2$ and consequently is equal to 2 or of the form $4n + 1$ because a prime number of the form $4n + 3$ is not the sum of two squares (cf. (2.13)).

Let q be a binary quadratic form, represented by the matrix M. A matrix $T \in GL(2, \mathbb{Z})$ is called a unit (automorph) of q if

$$T'MT = M,$$

i.e., if T maps the form q in itself. The unit T of q is called proper, if $\det(T) = 1$. We shall determine the proper units of a form q. Obviously they form a group $E(q)$. If the two forms q, q'—represented by matrices M, N, respectively—are properly equivalent, i.e.,

$$M = U'NU$$

with $U \in GL(2, \mathbb{Z})$, then the mapping $T \to UTU^{-1}$ yields an isomorphism $E(q) \cong E(q')$. Therefore, in view of (4.2), we may assume that q is reduced.

(4.5) **Theorem.** *The only proper units of $q(x, y) = a(x^2 + y^2)$ are $\pm \left(\begin{smallmatrix} 1 & 0 \\ 0 & 1 \end{smallmatrix} \right)$ and $\left(\begin{smallmatrix} 0 & \pm 1 \\ \mp 1 & 0 \end{smallmatrix} \right)$. The only proper units of $q(x, y) = a(x^2 + 2xy + y^2)$ are $\pm \left(\begin{smallmatrix} 1 & 0 \\ 0 & 1 \end{smallmatrix} \right)$, $\left(\begin{smallmatrix} 0 & \mp 1 \\ \pm 1 & +1 \end{smallmatrix} \right)$ and $\left(\begin{smallmatrix} \pm 1 & \pm 1 \\ \mp 1 & 0 \end{smallmatrix} \right)$. Any positive reduced form distinct from these two has only the trivial proper units $\pm \left(\begin{smallmatrix} 1 & 0 \\ 0 & 1 \end{smallmatrix} \right)$.*

PROOF. This proof follows from an analysis of the proof of the uniqueness statement in (4.2).

We have not yet discussed indefinite forms. One treats this case in a way similar to (4.2).

(4.6) **Theorem.** *An indefinite quadratic form is properly equivalent to a form with matrix $\left(\begin{smallmatrix} a & b \\ b & c \end{smallmatrix} \right)$ whose coefficients satisfy the following conditions:*

$$|a| \leqslant |c|, \qquad |b| \leqslant \frac{a}{2}.$$

(*In general the reduced form of an indefinite form is not uniquely determined.*)

The discriminant of an indefinite form is negative. This means that $\Delta = ac - b^2 < 0$ for the reduced form given by the above. Hence $ac < 0$ and $|\Delta| \geqslant 5b^2$, i.e.,

$$|b| \leqslant \sqrt{\frac{|\Delta|}{5}}$$

As in the positive case, one can write a table of reduced indefinite forms (Table 2). Among other things, the table shows that any odd natural number which is a divisor of $x^2 - 5y^2$ can be represented by $x^2 - 5y^2$ or $5x^2 - y^2$. In fact, there are three reduced forms with the same discriminant as $x^2 - 5y^2$, and the form $2x^2 + 2xy - 2y^2$ represents even numbers only.

Next Lagrange investigates the problem of representing prime numbers p by the form $x^2 + ay^2$, $a \in \mathbb{Z} - \{0\}$. He distinguishes between the cases $p = 4n - 1$ and $p = 4n + 1$.

Table 2

Δ	reduced indefinite forms (not necessarily inequivalent)
-1	$\begin{pmatrix} 1 & 0 \\ 0 & -1 \end{pmatrix}$
-2	$\begin{pmatrix} 1 & 0 \\ 0 & -2 \end{pmatrix}$ $\begin{pmatrix} -1 & 0 \\ 0 & 2 \end{pmatrix}$
-3	$\begin{pmatrix} 1 & 0 \\ 0 & -3 \end{pmatrix}$ $\begin{pmatrix} -1 & 0 \\ 0 & 3 \end{pmatrix}$
-4	$\begin{pmatrix} 1 & 0 \\ 0 & -4 \end{pmatrix}$ $\begin{pmatrix} -1 & 0 \\ 0 & 4 \end{pmatrix}$
-5	$\begin{pmatrix} 1 & 0 \\ 0 & -5 \end{pmatrix}$ $\begin{pmatrix} -1 & 0 \\ 0 & 5 \end{pmatrix}$ $\begin{pmatrix} 2 & 1 \\ 1 & -2 \end{pmatrix}$ $\begin{pmatrix} -2 & 1 \\ 1 & 2 \end{pmatrix}$
-6	$\begin{pmatrix} 1 & 0 \\ 0 & -6 \end{pmatrix}$ $\begin{pmatrix} -1 & 0 \\ 0 & 6 \end{pmatrix}$ $\begin{pmatrix} 2 & 0 \\ 0 & -3 \end{pmatrix}$ $\begin{pmatrix} -2 & 0 \\ 0 & 3 \end{pmatrix}$
-7	$\begin{pmatrix} 1 & 0 \\ 0 & -7 \end{pmatrix}$ $\begin{pmatrix} -1 & 0 \\ 0 & 7 \end{pmatrix}$ $\begin{pmatrix} 2 & 1 \\ 1 & -3 \end{pmatrix}$ $\begin{pmatrix} -2 & 1 \\ 1 & 3 \end{pmatrix}$
-8	$\begin{pmatrix} 2 & 0 \\ 0 & -4 \end{pmatrix}$ $\begin{pmatrix} 4 & 0 \\ 0 & -2 \end{pmatrix}$ $\begin{pmatrix} 1 & 0 \\ 0 & -8 \end{pmatrix}$ $\begin{pmatrix} -1 & 0 \\ 0 & 8 \end{pmatrix}$
-9	$\begin{pmatrix} 2 & 1 \\ 1 & -4 \end{pmatrix}$ $\begin{pmatrix} -2 & 1 \\ 1 & 4 \end{pmatrix}$ $\begin{pmatrix} 3 & 0 \\ 0 & -3 \end{pmatrix}$ $\begin{pmatrix} -3 & 0 \\ 0 & 3 \end{pmatrix}$
-10	$\begin{pmatrix} 1 & 0 \\ 0 & -10 \end{pmatrix}$ $\begin{pmatrix} -1 & 0 \\ 0 & 10 \end{pmatrix}$ $\begin{pmatrix} 2 & 0 \\ 0 & -5 \end{pmatrix}$ $\begin{pmatrix} -2 & 0 \\ 0 & 5 \end{pmatrix}$ $\begin{pmatrix} 3 & 1 \\ 1 & -3 \end{pmatrix}\begin{pmatrix} -3 & 1 \\ 1 & 3 \end{pmatrix}$

(4.7) **Theorem.** *Let a be an integer $\neq 0$. A prime number p of the form $p = 4n - 1$ is a divisor of $x^2 - ay^2$ if and only if p is not a divisor of $x^2 + ay^2$.*

PROOF. Let $p = 4n - 1$ be a divisor of $x^2 - ay^2$. Then in the field $\mathbb{F}_p$ with p elements, $x^2 - ay^2 = 0$, i.e., a is a square in $\mathbb{F}_p$. If in addition $x^2 + ay^2 = 0$ in $\mathbb{F}_p$, then $-a$ and consequently -1 would be a square in $\mathbb{F}_p$. We already saw (c.f. (2.13)) that this is not possible. Now let us assume that p is not a divisor of $x^2 - ay^2$. We have to show that in this case p is a divisor of $x^2 + ay^2$. It will suffice to show that $1 + a^{(p-1)/2} = 1^2 + a(a^{(p-3)/4})^2$ is a multiple of p $((p - 3)/4$ is an integer).

According to the so-called little theorem of Fermat, $a^{p-1} - 1 = (a^{(p-1)/2} - 1)(a^{(p-1)/2} + 1)$ is a multiple of p, i.e., it suffices to show that $a^{(p-1)/2} - 1$ is not a multiple of p. Let us assume that $a^{(p-1)/2} - 1$ is a multiple of p. Then the following polynomial identity

$$x^{p-1} - 1 = x^{p-1} - a^{(p-1)/2}$$

holds in $\mathbb{F}_p$. The last expression in this formula is a multiple of $x^2 - a$. This would imply that $x^2 - a$ decomposes into linear factors because (by Fermat's little theorem) $x^{p-1} - 1 = x^{p-1} - a(p-1)/2$ does. This means that a is a square in $\mathbb{F}_p$, i.e., there is $x_0 \in \mathbb{Z}$ such that p divides $x_0^2 - a$, i.e., is a factor of $x^2 - ay^2$, which is a contradiction.

Now we give a few typical applications which enable us to solve several of Fermat's problems (cf. (2.10)).

(4.8) Application. (1) Let p be a prime number of the form $p = 8n + 3$. Then p can be represented by the form $x^2 + 2y^2$. (2) Let p be a prime number of the form $p = 12n + 7$. Then p can be represented by $x^2 + 3y^2$. (3) Let p be a prime number of the form $p = 24n + 7$. Then p can be represented by $x^2 + 6y^2$.

PROOF. (1) Assume p is a divisor of $x^2 - 2y^2$. Then one can represent it by $x^2 - 2y^2$ or $-x^2 + 2y^2$ by (4.1) and Table 2. The only odd residues modulo 8 of these forms are ± 1 but not 3. Consequently, p is not a divisor of $x^2 - 2y^2$; by theorem (4.7) p is a divisor of $x^2 + 2y^2$, the only reduced form with discriminant 2. This means that $p = 8n + 3$ can be represented by $x^2 + 2y^2$.

(2) Assume $p = 12n + 7$ is a proper divisor of $x^2 - 3y^2$. Then it can be represented by $x^2 - 3y^2$ or $-x^2 + 3y^2$. However, these expressions have only ± 1, ± 9, ± 3 as their odd residues modulo 12, but not 7. Consequently, p is not a divisor of $x^2 - 3y^2$ which means that it is a divisor of $x^2 + 3y^2$ by (4.7). $2x^2 + 2xy + 2y^2$, the other reduced form of discriminant 3, represents even numbers only. Consequently, $x^2 + 3y^2$ represents $p = 12n + 7$.

(3) Assume $p = 24n + 7$ is a divisor of $x^2 + 6y^2$. Then it can be represented by $\pm(x^2 - 6y^2)$ or $\pm(2x^2 - 3y^2)$. One easily shows that these forms do not give the odd residue 7 modulo 24. According to (4.7) p is a divisor of $x^2 + 6y^2$; consequently, $p = 24n + 7$ can be represented by a form with determinant 6, i.e., by $x^2 + 6y^2$ or $2x^2 + 3y^2$ according to Table 7. This latter form does not leave the residue 7 modulo 24. Consequently, p can be represented by $x^2 + 6y^2$.

Let us now consider prime numbers p of the form $p = 4n + 1$. The following lemma is central.

(4.9) Lemma. p is a divisor of $x^2 + ay^2$ if and only if p is a divisor of $x^2 - ay^2$.

PROOF. p is a divisor of $x^2 \pm ay^2$ if and only if $x^2 \pm ay^2 = 0$ in $\mathbb{F}_p$. This means that $\mp a$ is a square in $\mathbb{F}_p$. For $p = 4n + 1$, -1 is a square in $\mathbb{F}_p$; consequently, $-a$ is a square in $\mathbb{F}_p$ if and only if a is a square in $\mathbb{F}_p$.

From this point on, Lagrange confines himself to prime numbers p of the form $p = 4an + 1$. One first observes that there is x_0 such that $x_0^{2an} + 1$ is a multiple of p because the following identity holds for the polynomial $x^{p-1} - 1$:

$$(x^{p-1} - 1) = (x^{(p-1)/2} + 1)(x^{(p-1)/2} - 1)$$

$$= (x^{2an} + 1)(x^{2an} - 1).$$

As we have seen above, $x^{p-1} - 1$ (and consequently $x^{2an} + 1$) decomposes into linear factors in $\mathbb{F}_p$. Also, x_0 can be chosen in such a way that $x_0^{2n} + 1$ is not a multiple of $p = 4an + 1$ for $a > 1$. Let us now set $y := x^n$, $z := y^2 + 1$. Then one has the remarkable identity

$$x^{2an} + 1 = z^a - az^{a-2}y^2 + \frac{a(a-3)}{2!} z^{a-4}y^4$$
$$- \frac{a(a-4)(a-5)}{3!} z^{a-6}y^6 + \frac{a(a-5)(a-6)(a-7)}{4!} z^{a-8}y^8$$
$$- + \cdots .$$

Since $y_0 = x_0^n$ and $z_0 = y_0^2 + 1$ are relatively prime, a consequence of this formula is that $p = 8n + 1$ is a divisor of the form $x^2 - 2y^2$ for $a = 2$. By (4.8) p is a divisor of $x^2 + 2y^2$, the only reduced form with discriminant 2. According to (4.1) p can be represented by this form.

When $a = 3$ we have the following consequences. p is a divisor of the number $z_0^3 - 3z_0y_0^2 = z_0(z_0^2 - 3y_0^2)$ and, since p and z_0 are relatively prime, also of $z_0^2 - 3y_0^2$. Consequently, $p = 12n + 1$ is a divisor of the form $x^2 - 3y^2$ and, according to (4.8), also of $x^2 + 3y^2$. Other than $x^2 + 3y^2$, only the reduced form $2x^2 + 2xy + 2y^2$ has discriminant 3. The latter does not have residue 1 modulo 12. This means that p can be represented by $x^2 + 3y^2$.

If $a = 5$ then $p = 2l + 1$ is a divisor of the number $z_0^5 - 5z_0^3y_0^2 + 5z_0y_0^4 = z_0(z_0^4 - 5z_0^2y_0^2 + 5y_0^4)$, i.e., of $z_0^4 - 5z_0^2y_0^2 + 5y_0^4$ and consequently also of $4z_0^4 - 20z_0^2y_0^2 + 20y_0^4 = (2z_0^2 - 5y_0^2)^2 - 5y_0^4$. This means that p is a divisor of the form $x^2 - 5y^2$ and consequently of $x^2 + 5y^2$ by (4.8). According to Table 1, p can be represented by $x^2 + 5y^2$ or $2x^2 + 2xy + 3y^2$. The latter equation does not have residue 1 modulo 20 which means that p is represented by $x^2 + 5y^2$. We summarize the above:

(4.10) **Application.** Prime numbers of the form $8n + 1$ can be represented by $x^2 + 2y^2$, prime numbers of the form $12n + 1$ can be represented by $x^2 + 3y^2$, and prime numbers of the form $20n + 1$ can be represented by $x^2 + 5y^2$.

It is not difficult to prove more theorems like this; see, for instance, Lagrange's paper.

The problem of representing numbers by quadratic forms inevitably leads to the solution of quadratic congruences (cf. (5.2)), i.e., to the law of quadratic reciprocity (cf. (5.1)). We will discuss this in the following chapter.

Next we look at Lagrange's solution of Fermat's equation

$$x^2 - dy^2 = 1.$$

Here Lagrange makes essential use of the theory of continued fractions which he substantially extended for this purpose. These solutions essentially coincide with the units of the quadratic diagonal form

$$\begin{pmatrix} 1 & 0 \\ 0 & -d \end{pmatrix};$$

for if

$$\begin{pmatrix} x & y \\ u & v \end{pmatrix}\begin{pmatrix} 1 & 0 \\ 0 & -d \end{pmatrix}\begin{pmatrix} x & u \\ y & v \end{pmatrix} = \begin{pmatrix} 1 & 0 \\ 0 & -d \end{pmatrix},$$

then by simple calculations,

$$x^2 - dy^2 = 1, \qquad xu - dyv = 0, \qquad u^2 - dv^2 = -d.$$

For $x = \pm 1$, $y = 0$ one obtains $u = 0$, $v = \pm 1$. For $x, y \neq 0$ one obtains

$$u = \frac{dyv}{x}, \qquad -d = \frac{d^2y^2v^2}{x^2} - dv^2.$$

This latter relation yields

$$dy^2v^2 - v^2x^2 = -x^2, \qquad v^2(x^2 - dy^2) = x^2, \qquad v^2 = x^2, \qquad v = \pm x$$

and consequently

$$u = \pm dy.$$

Therefore, the units are

$$\pm\begin{pmatrix} 1 & 0 \\ 0 & 1 \end{pmatrix}, \qquad \pm\begin{pmatrix} 1 & 0 \\ 0 & -1 \end{pmatrix}, \qquad \begin{pmatrix} x & dy \\ y & x \end{pmatrix}, \qquad \begin{pmatrix} x & -dy \\ y & -x \end{pmatrix}$$

and the units of determinant 1 are

$$\pm\begin{pmatrix} 1 & 0 \\ 0 & 1 \end{pmatrix}, \qquad \begin{pmatrix} x & dy \\ y & x \end{pmatrix}.$$

In particular, the set of solutions of Fermat's equation can be interpreted as a group in a natural way.

Lagrange uses the so-called continued fraction algorithm for the solution of Fermat's equation. After many individual results and more or less accidentally discovered connections, Euler and, even more so, Lagrange developed the theory of continued fractions in a systematic way. Euler is even more a member of the "naive" period of discovery, calculation and heuristic methods. But modern mathematics with its rigorous proofs, systematic procedures, and clear descriptions and delineations of the problems begins with Lagrange. A decisive change took place in the development of number theory between Euler and Lagrange.

We will now describe the theory of continued fractions. As references we mention Niven and Zuckermann, *An Introduction to the Theory of Numbers*, Hardy and Wright, and Hasse.

Let $[\theta]$ denote the largest integer $\leq \theta$ for $\theta \in \mathbb{R}$. $[\ ]$ is called the Gauss bracket (or the "greatest integer function.") For $\theta \notin \mathbb{Z}$ we define

$$\theta := a_0 + \frac{1}{\theta_1} \qquad \text{with} \quad a_0 := [\theta], \quad \theta_1 > 1.$$

We continue in this way:

$$\theta_1 := a_1 + \frac{1}{\theta_2} \qquad \text{with} \quad a_1 := [\theta_1], \quad \theta_2 > 1, \qquad \text{if} \quad \theta_1 \notin \mathbb{Z}$$

$$\theta_n := a_n + \frac{1}{\theta_{n+1}} \qquad \text{with} \quad a_n := [\theta_n], \quad \theta_{n+1} > 1, \quad \text{if} \quad \theta_n \notin \mathbb{Z}.$$

This definition leads to

$$\theta = a_0 + \cfrac{1}{a_1 + \cfrac{1}{a_2 + \cfrac{\ddots}{\quad + a_n + \cfrac{1}{\theta_{n+1}}}}}\,.$$

The sequence $a_0, a_1, a_2, \ldots$ is called the *expansion* of θ into a *continued fraction*.

(4.11) **Remark.** The expansion into a continued fraction terminates if and only if θ is a rational number.

PROOF. If the expansion breaks off, then $\theta_n = a_n$ is an integer. Then

$$\theta = a_0 + \cfrac{1}{a_1 + \cfrac{\ddots}{a_{n-1} + \cfrac{1}{a_n}}}$$

which is obviously rational. Conversely, if $\theta = u/v$ is rational we use the Euclidean algorithm to write

$$u = a_0 v + r_1, \qquad 0 < r_1 < v,$$
$$v = a_1 r_1 + r_2, \qquad 0 < r_2 < r_1,$$
$$r_1 = a_2 r_2 + r_3, \qquad 0 < r_3 < r_2,$$
$$\cdots .$$

This procedure breaks off after finitely many steps with $r_{n-1} = a_n r_n$. The equations are equivalent to

$$\theta = a_0 + \frac{r_1}{v} = a_0 + \frac{1}{\theta_1},$$
$$\theta_1 = a_1 + \frac{r_2}{r_1} = a_1 + \frac{1}{\theta_2},$$
$$\theta_2 = a_2 + \frac{r_3}{r_2} = a_2 + \frac{1}{\theta_3},$$
$$\cdots .$$

Eventually one obtains $\theta_n = r_{n-1}/r_n \in \mathbb{Z}$, and the expansion into a continued fraction ends.

It is interesting to find the continuous fraction expansions of a few numbers:

π:	3	7	15	1	293	$\ldots$				
e:	2	1	2	1	1	4	1	1	6	1 $\ldots$
$\sqrt{2}$:	1	2	2	2	$\ldots$					
$\sqrt{3}$:	1	1	2	1	2	1	$\ldots$			
$\sqrt{5}$:	2	4	4	4	$\ldots$					
$\sqrt{6}$:	2	2	4	2	4	2	$\ldots$			

(for e, see A. Hurwitz, Über die Kettenbruchentwicklung der Zahl e, *Gesammelte Abhandlungen* II; Euler showed that generally $\sqrt{n^2+1} = n, 2n, \ldots$.)

Quadratic irrationalities have periodic expansions, a phenomenon which will be investigated later.

Let us now derive a number of formulas that will be needed later. For $a_0, a_1, \ldots, a_n \in \mathbb{R}$, $a_1, \ldots, a_n \geqslant 1$, we define

$$\langle a_0, a_1, \ldots, a_n \rangle := a_0 + \cfrac{1}{a_1 + \cfrac{\cdot \cdot}{\quad a_{n-1} + \cfrac{1}{a_n}}}.$$

For $a_n > 1$ one has $a_n = a_n - 1 + 1/1$ and consequently

$$\langle a_0, a_1, \ldots, a_n \rangle = \langle a_0, a_1, \ldots, a_n - 1, 1 \rangle.$$

The expansion of a rational number as a continued fraction is unique except for this identity.

(4.12) **Remark.** Let $\langle a_0, \ldots, a_m \rangle = \langle b_0, \ldots, b_n \rangle$ with $a_i, b_i \in \mathbb{Z}$, $a_1, \ldots, b_1, \ldots \geqslant 1$ and $a_m, b_n > 1$. Then $m = n$ and $a_i = b_i$ for all i.

PROOF. The proof follows by induction from

$$\langle a_0, \ldots, a_m \rangle = a_0 + \frac{1}{\langle a_1, \ldots, a_m \rangle} = b_0 + \frac{1}{\langle b_1, \ldots, b_n \rangle}$$

if we can only show that $\langle a_1, \ldots, a_m \rangle > 1$ whenever $a_1, \ldots, a_m \geqslant 1$, $a_m > 1$. But this is obvious from

$$\langle a_1, \ldots, a_m \rangle = a_1 + \cfrac{1}{a_2 + \cfrac{\cdot\cdot}{\cdot}}$$

Let $a_0, a_1, a_2, \ldots$ be a sequence with $a_1, a_2, \ldots > 0$. Let

$$\tau_n := \langle a_0, a_1, a_2, \ldots, a_n \rangle.$$

τ_n can be computed with the help of the formulas

$$p_0 = a_0, \qquad p_1 = a_0 a_1 + 1, \ldots, \qquad p_n = a_n p_{n-1} + p_{n-2},$$
$$q_0 = 1, \qquad q_1 = a_1, \ldots, \qquad q_n = a_n q_{n-1} + q_{n-2}. \tag{4.13}$$

Then

$$\frac{p_0}{q_0} = \frac{a_0}{1} = \tau_0, \qquad \frac{p_1}{q_1} = a_0 + \frac{1}{a_1} = \tau_1, \ldots$$

and more generally:

(4.14) Remark. $p_n / q_n = \tau_n$, specifically,

$$\theta = \frac{\theta_n p_{n-1} + p_{n-2}}{\theta_n q_{n-1} + q_{n-2}}.$$

PROOF. This is proved by induction. The cases $n = 0, 1$ have already been discussed. Suppose that

$$\tau_n = \langle a_0, a_1, \ldots, a_n \rangle = \left\langle a_0, a_1, \ldots, a_{n-1} + \frac{1}{a_n} \right\rangle$$

$$= \frac{p'_{n-1}}{q'_{n-1}},$$

where p'_{n-1}, q'_{n-1} are the p, q belonging to $a_0, \ldots, a_{n-2}, a_{n-1} + 1/a_n$. Then

$$\frac{p'_{n-1}}{q'_{n-1}} = \frac{(a_{n-1} + 1/a_n) p_{n-2} + p_{n-3}}{(a_{n-1} + 1/a_n) q_{n-2} + q_{n-3}} = \frac{p_{n-1} + (1/a_n) p_{n-2}}{q_{n-1} + (1/a_n) q_{n-2}}.$$

The second statement follows from the first and $\theta = \langle a_0, \ldots, a_{n-1}, \theta_n \rangle$.

τ_n is called the nth convergent to the sequence $a_0, a_1, a_2, \ldots$.

(4.15) Theorem. *Let $a_0 \in \mathbb{Z}$; $a_1, a_2, \ldots \in \mathbb{N}$. Then the sequence $\{\tau_n\}_{n=1,2,\ldots}$ converges to θ, where θ is an irrational number. The a_i are uniquely defined by the expansion of θ as a continued fraction. Conversely, let θ be an arbitrary irrational number. Then $\theta = \lim \tau_n$ if $\tau_n = \langle a_0, \ldots, a_n \rangle$ is obtained by expanding θ as a continued fraction.*

PROOF. Obviously, $\{p_i\}, \{q_i\}$ are strictly monotonically increasing sequences of natural numbers. Convergence follows from

$$\tau_n - \tau_{n-1} = (-1)^{n-1} \frac{1}{q_n q_{n-1}} \tag{4.16}$$

because this formula shows that the differences form an alternating sequence converging to 0. Relation (4.16) is equivalent to

$$p_n q_{n-1} p_{n-1} q_n = (-1)^{n-1}, \tag{4.16}'$$

and this formula follows trivially by induction.

With these considerations we have nearly completed the proof of the second statement. Let $a_0, a_1, \ldots$ be the expansion of θ in a continued fraction. Let $\tau_0, \tau_1, \ldots$ be the convergents. If we apply (4.16) to

$$\theta = \langle a_0, a_1, \ldots, a_{n-1}, \theta_n \rangle$$

we obtain

$$\theta - \tau_{n-1} = (-1)^{n-1} \frac{1}{q_{n-1}(\theta_n q_{n-1} + q_{n-2})}$$

Since $\theta_n > 0$, $q_i \to +\infty$, it follows that $\lim_n \tau_n = \theta$. It remains to show the uniqueness of the expansion in a continued fraction. This is done analogously to the rational case (4.12).

A continued fraction of the form

$$\langle a_0, \ldots, a_{n-1}, b_1, \ldots, b_k, b_1, \ldots, b_k, \ldots \rangle$$

is called periodic. Occasionally we use the abbreviation

$$\langle a_0, \ldots, a_{n-1}, \overline{b_1, \ldots, b_k} \rangle.$$

$a_0, \ldots, a_{n-1}$ is called the "preperiod," $b_1, \ldots, b_k$ the "period."

(4.17) Theorem. θ *can be expanded into a periodic continued fraction if and only if it is of the form* $\alpha + \beta\sqrt{d}$ *with* $\alpha, \beta \in \mathbb{Q}$ *and* $d \in \mathbb{N}$ *not a square.*

PROOF. Let us assume we have a periodic continued fraction. First we consider the purely periodic case

$$\theta = \langle a_0, a_1, \ldots, a_n, a_0, a_1, \ldots, a_n, \ldots \rangle.$$

Then

$$\theta = \langle a_0, a_1, \ldots, a_n, \theta \rangle,$$

hence

$$\theta = \frac{\theta p_{n-1} + p_{n-2}}{\theta q_{n-1} + q_{n-2}}.$$

This is a quadratic equation for θ. If there is a preperiod, say

$$\theta = \langle a_0, \ldots, a_m, \overline{b_1, \ldots, b_n}, \ldots \rangle,$$

then we set

$$\tau := \langle \overline{b_1, \ldots, b_n} \rangle, \qquad \theta := \langle a_0, \ldots, a_m, \tau \rangle, \qquad \theta = \frac{\tau p_m + p_{m-1}}{\tau q_m + q_{m-1}}.$$

So there is a rational relation between θ and τ, i.e., θ is of the form $\alpha + \beta\sqrt{d}$.

Conversely, a quadratic irrationality θ can be written as

$$\theta = \frac{a + \sqrt{b}}{c} = \frac{ac + \sqrt{bc^2}}{c^2} \quad \text{or} \quad = \frac{-ac + \sqrt{bc^2}}{-c^2} \qquad (a, b, c \in \mathbb{Z})$$

$$= \frac{m_0 + \sqrt{d}}{k_0} \qquad (m_0, k_0, d \in \mathbb{Z}),$$

where k_0 is a factor of $m_0^2 - d$. We recursively define

$$\theta := \theta_0, \qquad \theta_i := \frac{m_i + \sqrt{d}}{k_i}, \qquad a_i := [\theta_i],$$

$$m_{i+1} := a_i k_i - m_i, \qquad k_{i+1} := \frac{d - m_{i+1}^2}{k_i}.$$ (4.18)

We will now show that $a_i, m_i, k_i \in \mathbb{Z}$ and that the a_i give the expansion of $\theta = \theta_0$ as a continued fraction. Clearly, a_i, m_i, k_i are sequences of real numbers. $m_i, k_i \in \mathbb{Z}$ for $i = 0$, and k_i is a factor of $d - m_i^2$. Let us assume that this holds for i. Then for $i + 1$ we obtain

$$m_{i+1} \in \mathbb{Z}, \qquad k_{i+1} = \frac{d - a_i^2 k_i^2 + 2a_i k_i m_i - m_i^2}{k_i}$$

$$= \frac{d - m_i^2}{k_i} + 2a_i m_i - a_i^2 k_i \in \mathbb{Z}.$$

From $k_i = (d - m_{i+1}^2)/k_{i+1}$, it follows that $k_i + 1$ is a divisor of $d - m_{i+1}^2$. We obtain directly from the definitions

$$\theta_i - a_i = \frac{-a_i k_i + m_i + \sqrt{d}}{k_i} = \frac{\sqrt{d} - m_{i+1}}{k_i}$$

$$= \frac{d - m_{i+1}^2}{k_i \sqrt{d} + k_i m_{i+1}} = \frac{k_{i+1}}{m_{i+1} + \sqrt{d}} = \frac{1}{\theta_{i+1}}.$$

Hence $\theta = \langle a_0, a_1, a_2, \ldots \rangle$.

To this point we have been considering a slightly different way of expanding a number as a continued fraction. We will now start with the proof proper. Let $\xi' = \alpha - \beta\sqrt{d}$ be the conjugate of $\xi = \alpha + \beta\sqrt{d}$; then $(\xi + \eta)' = \xi' + \eta'$, $(\xi\eta)' = \xi'\eta'$, $(\xi/\eta)' = \xi'/\eta'$. We obtain

$$\theta' = \theta_0' = \frac{\theta_n' p_{n-1} + p_{n-2}}{\theta_n' q_{n-1} + q_{n-2}}$$

and solve this equation for θ_n':

$$\theta_n' = -\frac{q_{n-2}}{q_{n-1}} \left(\frac{\theta_0' - p_{n-2}/q_{n-2}}{\theta_0' - p_{n-1}/q_{n-1}} \right).$$

The number in the parenthesis converges to $1 = (\theta_0' - \theta)/(\theta_0' - \theta)$ for $n \to \infty$. Hence $\theta_n' < 0$ for $n > N_0$. Because $\theta_n > 0$ we also have $\theta_n - \theta_n' > 0$.

By the definition of θ_n,

$$\theta_n - \theta_n' = \frac{2\sqrt{d}}{k_n} > 0,$$

and specifically

$$0 < k_n.$$

From (4.18) it follows that

$$0 < k_n k_{n+1} = d - m_{n+1}^2 \leqq d,$$

$$m_{n+1}^2 < m_{n+1}^2 + k_n k_{n+1} \leqq d, \qquad |m_{n+1}| < \sqrt{d}.$$

Consequently, for $n > N_0$, the numbers k_n, m_n will assume only finitely many values. Thus there are indices $n < j$ with $k_n = k_j$, $m_n = m_j$, consequently $\theta_n = \theta_j$, and hence

$$\theta = \langle a_0, \ldots, a_{n-1}, \overline{a_n, a_{n+1}, \ldots, a_{j-1}} \rangle.$$

(4.19) Theorem. *The continued fraction of a quadratic irrationality θ is purely periodic if and only if $1 < \theta$, $-1 < \theta' < 0$.*

PROOF. We start out by assuming that $1 < \theta$ and $-1 < \theta' < 0$. Then

$$\theta_i = a_i + \frac{1}{\theta_{i+1}}, \qquad \frac{1}{\theta_{i+1}'} = \theta_i' - a_i.$$

We know that $a_i \geqslant 1$ for every i, and also for $i = 0$, since $\theta > 1$. For $\theta_i' < 0$ one obtains $1/\theta_{i+1}' < -1$ and $-1 < \theta_{i+1}' < 0$. Since $-1 < \theta' < 0$, it follows by induction that $-1 < \theta_i' < 0$ for all $i \geqslant 0$. Hence

$$0 < -\frac{1}{\theta_{i+1}'} - a_i < 1, \qquad a_i = \left[-\frac{1}{\theta_{i+1}'} \right].$$

As θ is a quadratic irrationality, its continued fraction is periodic. This means that there are indices $n < j$ with $\theta_n = \theta_j$, $a_n = a_j$. Then it follows that $-1/\theta_n' = -1/\theta_j'$, $[-1/\theta_n'] = [-1/\theta_j']$ and $a_{n-1} = a_{j-1}$. Induction shows that the fraction is purely periodic.

Conversely, let us assume that we have a purely periodic expansion of θ into a continued fraction, say

$$\theta = \langle \overline{a_0, a_1, \ldots, a_{n-1}} \rangle \qquad (a_i \in \mathbb{Z}, a_i > 0).$$

Then, because $\theta > a_0 \geqslant 1$ and

$$\theta = \frac{\theta_n P_{n-1} + P_{n-2}}{\theta_n q_{n-1} + q_{n-2}}$$

we have

$$\theta = \langle a_0, a_1, \ldots, a_{n-1}, \theta \rangle = \frac{\theta p_{n-1} + p_{n-2}}{\theta q_{n-1} + q_{n-2}}.$$

This means that θ satisfies the quadratic equation

$$f(x) = x^2 q_{n-1} + x(q_{n-2} - p_{n-1}) - p_{n-2} = 0.$$

θ and θ' are the solutions of this equation. Since $\theta > 1$ it suffices to show that the equation has a root between -1 and 0. To show this, it suffices to show that $f(-1)$ and $f(0)$ have opposite signs. By the definition of p_n, $f(0) = -p_{n-2} < 0$. Moreover, for $n > 1$, we have

$$f(-1) = q_{n-1} - q_{n-2} + p_{n-1} - p_{n-2}$$
$$= (q_{n-2} + p_{n-2})(a_{n-1} - 1) + q_{n-3} + p_{n-3}$$
$$\geqq q_{n-3} + p_{n-3} > 0.$$

Finally, for $n = 1$ we have $f(-1) = a_0 > 0$.

Let us now expand $\sqrt{d}$ as a continued fraction where d is a positive integer that is not a square. We consider the irrational number $\theta := \sqrt{d} + [\sqrt{d}]$. Then $\theta > 1$, $\theta' = -\sqrt{d} + [\sqrt{d}]$ and $-1 < \theta' < 0$. By Theorem (4.19), θ has a purely periodic expansion as a continued fraction, say

$$\theta = \langle \overline{a_0, \ldots, a_{n-1}} \rangle = \langle a_0, \overline{a_1, \ldots, a_{n-1}, a_0} \rangle.$$

Let n be its minimal period. $\theta_i := \langle a_i, a_{i+1}, \ldots \rangle$ is purely periodic for all i and $\theta = \theta_0 = \theta_n = \theta_{2n} = \cdots . \theta_0, \theta_1, \ldots, \theta_{n-1}$ are all different, otherwise n would not be the minimal length of the period. Then $\theta_i = \theta_0$ if and only if $i = nj$ for a certain j. By (4.18) on page 49 we start out with $\theta_0 = (m_0 + \sqrt{d})$, $k_0 = 1$, $m_0 = [\sqrt{d}]$. Then

$$\frac{m_{nj} + \sqrt{d}}{k_{nj}} = \theta_{nj} = \theta_0 = \frac{m_0 + \sqrt{d}}{k_0}$$

$$= [\sqrt{d}] + \sqrt{d}$$

and consequently

$$m_{nj} - k_{nj}[\sqrt{d}] = (k_{nj} - 1)\sqrt{d}.$$

This means that $k_{nj} = 1$ because the left-hand side is rational and $\sqrt{d}$ irrational. We claim $k_i = 1$ only for $i = nj$. In fact, $\theta_i = m_i + \sqrt{d}$ follows from $k_i = 1$. However, θ_i has a purely periodic expansion as a continued fraction, and one has, by Theorem (4.19), $-1 < m_i - \sqrt{d} < 0$, $\sqrt{d} - 1 < m_i < \sqrt{d}$, and consequently $m_i = [\sqrt{d}]$, i.e., $\theta_i = \theta_0$, and i is a multiple of n. In fact, $k_i \neq -1$ for every i, for $k_i = -1$ implies $\theta_i = -m_i - \sqrt{d}$ and $-m_i - \sqrt{d} > 1$ and $-1 < -m_i + \sqrt{d} < 0$ according to Theorem (4.19). This leads to the contradiction $\sqrt{d} < m_i < -\sqrt{d} - 1$. Since $a_0 = [\sqrt{d} + [\sqrt{d}]] = 2[\sqrt{d}]$, one has

$$\sqrt{d} = -[\sqrt{d}] + (\sqrt{d} + [\sqrt{d}])$$

$$= -[\sqrt{d}] + \langle 2[\sqrt{d}], \overline{a_1, a_2, \ldots, a_{n-1}, a_0} \rangle$$

$$= \langle [\sqrt{d}], \overline{a_1, a_2, \ldots, a_{n-1}, a_0} \rangle$$

with $a_0 = 2[\sqrt{d}]$.

Applying the formulas (4.18) to $\sqrt{d} + [\sqrt{d}\,]$, $k_0 = 1$, $m_0 = [\sqrt{d}\,]$ leads to $a_0 = 2[\sqrt{d}\,]$, $m_1 = [\sqrt{d}\,]$, $k_1 = d - [\sqrt{d}\,]^2$. If one applies these formulas to $\sqrt{d}$ with $k_0 = 1$, $m_0 = 0$, one obtains $a_0 = [\sqrt{d}\,]$, $m_1 = [\sqrt{d}\,]$, $k_1 = d - [\sqrt{d}\,]^2$. This means that a_0 assumes different values but m_1 and k_1 remain constant. Since $\theta_i = (m_i + \sqrt{d}\,)/k_i$, these formulas yield the same values for a_i, m_i, and k_i $(i \neq 0)$. We see that the expansions as continued fractions for $\sqrt{d} + [\sqrt{d}\,]$ and $\sqrt{d}$ differ only in a_0 and m_0.

We are now in a position to solve the equation

$$x^2 - dy^2 = \pm 1, \qquad d \text{ not a square}.$$

Let us first state:

(4.20) **Theorem.** *Using the definitions above, we have* $p_{n-1}^2 - q_{n-1}^2 d = (-1)^n k_n$.

PROOF. Let us begin with the identity

$$\theta = \sqrt{d} = \frac{\theta_n p_{n-1} + p_{n-2}}{\theta_n q_{n-1} + q_{n-2}} = \frac{\left((m_n + \sqrt{d}\,)/k_n\right)p_{n-1} + p_{n-2}}{\left((m_n + \sqrt{d}\,)/k_n\right)q_{n-1} + q_{n-2}}$$

$$= \frac{m_n p_{n-1} + \sqrt{d}\, p_{n-1} + k_n p_{n-2}}{m_n q_{n-1} + \sqrt{d}\, q_{n-1} + k_n q_{n-2}}.$$

If one multiplies this equation by its denominator and separates the rational from the irrational part, it follows that

$$dq_{n-1} = m_n p_{n-1} + k_n p_{n-2},$$

$$p_{n-1} = m_n q_{n-1} + k_n q_{n-2}.$$

Multiplying the first equation by q_{n-1} and the second by p_{n-1} and subtracting the second from the first equation yields

$$dq_{n-1}^2 - p_{n-1}^2 = k_n (p_{n-2} q_{n-1} - q_{n-2} p_{n-1}).$$

By (4.16)', $p_{n-2} q_{n-1} - q_{n-2} p_{n-1} = (-1)^{n-1}$.

Let n be the length of the period of the expansion of $\sqrt{d}$ as a continued fraction. Then $k_n = 1$, and the following corollary holds for every $j \in \mathbb{N}$.

(4.21) **Corollary.** $p_{nj-1}^2 - dq_{nj-1}^2 = (-1)^{nj}$.

(4.22) **Corollary.** *The equation* $x^2 - dy^2 = 1$ *has infinitely many solutions. For n even, $x = p_{nj-1}$, $y = q_{nj-1}$, and for n odd, $x = p_{2nj-1}$, $y = q_{2nj-1}$. If n is odd the equation $x^2 - dy^2 = -1$ has infinitely many solutions of the form $x = p_{nj-1}$, $y = q_{nj-1}$ for odd j.*

The next theorem shows that every solution of $x^2 - dy^2 = \pm 1$ is obtained from the expansion of $\sqrt{d}$ as a continued fraction.

First we make the following simple observation. Save for the trivial solutions $x = \pm 1$, $y = 0$ of $x^2 - dy^2 = 1$ (and, analogously, the corresponding trivial solutions of $x^2 - dy^2 = n$), every solution of $x^2 - dy^2 = n$ yields three additional solutions, by combining all the possible signs of $\pm x$ and $\pm y$. Hence, it suffices to consider positive solutions $x > 0$, $y > 0$.

(4.23) **Theorem.** *Let d be a natural number which is not a square and p_n/q_n the convergents in the expansion of $\sqrt{d}$ as a continued fraction. If $|N| < \sqrt{d}$ and s, t is a solution of $x^2 - dy^2 = N$ in natural numbers such that g.c.d.(s, t) $= 1$, then there is an index n such that $s = p_n$, $t = q_n$.*

PROOF. Let E and M be natural numbers with g.c.d.$(E, M) = 1$ and $E^2 - \rho M^2 = \sigma$, where $\sqrt{\sigma}$ is irrational and $0 < \sigma < \sqrt{\rho}$, $\sigma, \rho \in \mathbb{R}$. Then

$$\frac{E}{M} - \sqrt{\rho} = \frac{\sigma}{M(E + M\sqrt{\rho})},$$

and consequently

$$0 < \frac{E}{M} - \sqrt{\rho} < \frac{\sqrt{\rho}}{M(E + M\sqrt{\rho})} = \frac{1}{M^2((E/M\sqrt{\rho}) + 1)}.$$

From $0 < E/M - \sqrt{\rho}$ it follows that $E/M\sqrt{\rho} > 1$. Consequently

$$\left| \frac{E}{M} - \sqrt{\rho} \right| < \frac{1}{2M^2}.$$

According to lemma (4.24) which we will prove presently, E/M is a convergent to the continued fraction expansion of $\sqrt{\rho}$.

(4.24) **Lemma.** *Let θ be an arbitrary irrational number. Assume the rational number a/b satisfies*

$$\left| \theta - \frac{a}{b} \right| < \frac{1}{2b^2} \qquad \text{with} \quad b \geqslant 1.$$

Then a/b is a convergent in the expansion of θ as a continued fraction.

We now complete the proof of (4.23). When $N > 0$ we set $\sigma = N$, $\rho = d$, $E = s$, $M = t$, and the theorem is obviously true. If $N < 0$ then $t^2 - (1/d)s^2$ $= -N/d$. We set $\sigma = -N/d$, $\rho = 1/d$, $E = t$, $M = s$. One can easily see that t/s is a convergent in the expansion of $1/\sqrt{d}$. By the lemma below, s/t is a convergent in the expansion of $\sqrt{d}$.

(4.25) **Lemma.** *The nth convergent to $1/x$ is the reciprocal of the $(n-1)$th convergent to x for $x \in \mathbb{R}$, $x > 1$.*

We can summarize our results as follows:

(4.26) Theorem. *All positive solutions of $x^2 - dy^2 = \pm 1$ are given by the convergents of the expansion of $\sqrt{d}$ as a continued fraction. Let n be the length of the period of the expansion of $\sqrt{d}$ and even. Then $x^2 - dy^2 = -1$ does not have a solution; all positive solutions of $x^2 - dy^2 = 1$ are given by $x = p_{nj-1}$, $y = q_{nj-1}$ for $j = 1, 2, \ldots$. For n odd, all positive solutions of $x^2 - dy^2 = -1$ are given by $x = p_{nj-1}$ and $y = q_{nj-1}$ for $j = 1, 3, 5 \ldots$. These are all the positive solutions of $x^2 = dy^2 - 1$ for $j = 2, 4, 6, \ldots$.*

The sequence of the pairs $(p_0, q_0), (p_1, q_1), \ldots$ contains all positive solutions of $x^2 - dy^2 = 1$. Since $a_0 = [\sqrt{d}] > 0$, the sequence $p_0, p_1, p_2, \ldots$ is strictly monotonically increasing. Let x_1, y_1 be the first solution. For all other solutions, $x > x_1$ and $y > y_1$. After finding the smallest positive solution with the help of the continued fraction, one can find all other positive solutions by a simple method:

(4.27) Theorem. *Let x_1, y_1 be the minimal solution of $x^2 - dy^2 = 1$ in natural numbers ($d > 0$, not a square). All further solutions in natural numbers are given by x_n, y_n, $n = 1, 2, 3, \ldots$ with x_n, y_n defined by*

$$x_n + y_n\sqrt{d} = \left(x_1 + y_1\sqrt{d}\right)^n.$$

(One computes x_n and y_n by expanding the right-hand side into a rational and irrational part.)

PROOF. Obviously, $x_n - y_n\sqrt{d} = (x_1 - y_1\sqrt{d})^n$. Hence

$$x_n^2 - y_n^2 d = \left(x_n - y_n\sqrt{d}\right)\left(x_n + y_n\sqrt{d}\right)$$
$$= \left(x_1 - y_1\sqrt{d}\right)^n\left(x_1 + y_1\sqrt{d}\right)^n$$
$$= \left(x_1^2 - y_1^2 d\right)^n$$
$$= 1.$$

Every positive solution is obtained in this way. In fact, assume that there is a positive solution (s, t) which does not correspond to any of the (x_n, y_n). Since $x_1 + y_1\sqrt{d} > 1$ and $s + t\sqrt{d} > 1$ there is an m such that

$$\left(x_1 + y_1\sqrt{d}\right)^m \leqslant s + t\sqrt{d} < \left(x_1 + y_1\sqrt{d}\right)^{m+1}.$$

$(x_1 + y_1\sqrt{d})^m = s + t\sqrt{d}$ is impossible because this leads to $x_m + y_m\sqrt{d} = s + t\sqrt{d}$, i.e., $s = x_m$, $t = y_m$. Clearly $(x_1 - y_1\sqrt{d})^m = (x_1 + y_1\sqrt{d})^{-m}$. Multiplying the above inequality by $(x_1 - y_1\sqrt{d})^m$, one obtains

$$1 < \left(s + t\sqrt{d}\right)\left(x_1 - y_1\sqrt{d}\right)^m < x_1 + y_1\sqrt{d}.$$

We define the integers a, b by

$$\left(a + b\sqrt{d}\,\right) = \left(s + t\sqrt{d}\,\right)\left(x_1 - y_1\sqrt{d}\,\right)^m$$

and obtain

$$a^2 - b^2 d = (s^2 - t^2 d)\left(x_1^2 - y_1^2 d\right)^m = 1$$

This means that a, b is the solution of $x^2 - dy^2 = 1$ with $1 < a + b\sqrt{d}$ $< x_1 + y_1\sqrt{d}$. On the other hand, $0 < (a + b\sqrt{d}\,)^{-1} < 1$, i.e., $0 < a - b\sqrt{d}$ < 1. So

$$a = \tfrac{1}{2}\left(a + b\sqrt{d}\,\right) + \tfrac{1}{2}\left(a - b\sqrt{d}\,\right) > \tfrac{1}{2} + 0 > 0,$$

$$b\sqrt{d} = \tfrac{1}{2}\left(a + b\sqrt{d}\,\right) - \tfrac{1}{2}\left(a - b\sqrt{d}\,\right) > \tfrac{1}{2} - \tfrac{1}{2} = 0;$$

This means that a, b is a positive solution. Hence $a > x_1$, $b > y_1$ which contradicts $a + b\sqrt{d} < x_1 + y_1\sqrt{d}$.

We will now prove (4.24) and (4.25).
(4.25) is easy. Obviously, we have $x = \langle a_0, a_1, \ldots \rangle$ and $1/x = \langle 0, a_0, a_1, \ldots \rangle$. Let p_n/q_n and p_n'/q_n' be the convergents to x and $1/x$. Then

$$p_0' = 0, \qquad p_1' = 1, \qquad p_2' = a_1, \qquad p_n' = a_{n-1}p_{n-1}' + p_{n-2}',$$

$$q_0 = 1, \qquad q_1 = a_1, \qquad q_{n-1} = a_{n-1}q_{n-2} + q_{n-3},$$

$$q_0' = 1, \qquad q_1' = a_0, \qquad q_2' = a_0 a_1 + 1, \qquad q_n' = a_{n-1}q_{n-1}' + q_{n-2}',$$

$$p_0 = a_0, \qquad p_1 = a_0 a_1 + 1, \qquad p_{n-1} = a_{n-1}p_{n-2} + p_{n-3}.$$

The statement follows by induction.

Now we consider (4.24). Let g.c.d.$(a, b) = 1$ which we can assume without loss of generality. Let p_n/q_n be the convergents to θ. Suppose a/b is not one of them. Then the inequality $q_m \leqslant b < q_{m+1}$ defines a number m. We claim that $|\theta b - a| < |\theta q_m - p_m|$ is impossible. Suppose not. We consider the linear system of equations

$$q_m x + q_{m+1} y = b,$$

$$p_m x + p_{m+1} y = a.$$

We know (see (4.16)′) that the determinant of this system is $q_m p_{m+1} - q_{m+1}p_m = \pm 1$. Then an integral solution x, y exists with $x \neq 0$ and $y \neq 0$. For if $x = 0$, then $b = yq_{m+1}$ and hence $y > 0$ and $b \geqslant q_{m+1}$, which contradicts $b < q_{m+1}$. If $y = 0$ then $a = xp_m$, $b = xq_m$, and

$$|\theta b - a| = |\theta x q_m - x p_m| = |x|\,|\theta q_m - p_m| \geqq |\theta q_m - p_m|,$$

a contradiction because of $|x| \geqslant 1$.
x and y have different signs. If $y < 0$ then $x > 0$ follows immediately from $xq_m = b - yq_{m+1}$. If $y > 0$ then $b < yq_{m+1}$ follows immediately because $b < q_{m+1}$. In other words, xq_m is negative, and hence $x < 0$. Formu-

las (4.16) and (4.16)' show that $q_m - p_m$ and $q_{m+1} - p_{m+1}$ have different signs; consequently, $x(q_m - p_m)$ and $y(q_{m+1} - p_{m+1})$ have the same sign. From the equations which define x and y we have

$$\theta b - a = x(\theta q_m - p_m) + y(\theta q_{m+1} - p_{m+1}).$$

Since the two expressions on the right-hand side have the same sign, we have

$$
\begin{aligned}
|\theta b - a| &= |x(\theta q_m - p_m) + y(\theta q_{m+1} - p_{m+1})| \\
&= |x(\theta q_m - p_m)| + |y(\theta q_{m+1} - p_{m+1})| \\
&> |x(\theta q_m - p_m)| \\
&= |x| |\theta q_m - p_m| \\
&\geqq |\theta q_m - p_m|,
\end{aligned}
$$

a contradiction. This leads to

$$|\theta q_m - p_m| \leqq |\theta b - a| < \frac{1}{2b},$$

$$\left| \theta - \frac{p_m}{q_m} \right| < \frac{1}{2bq_m}.$$

Since $a/b \neq p_m/q_m$ we have

$$
\begin{aligned}
\frac{1}{bq_m} &\leqq \frac{|bp_m - aq_m|}{bq_m} = \left| \frac{p_m}{q_m} - \frac{a}{b} \right| \\
&\leqq \left| \theta - \frac{p_m}{q_m} \right| + \left| \theta - \frac{a}{b} \right| \\
&< \frac{1}{2bq_m} + \frac{1}{2b^2}.
\end{aligned}
$$

and hence $b < q_m$, a contradiction.

This concludes our discussion of continued fractions and our chapter on Lagrange. Let us end with a quote from Dirichlet:

This gap [the fact that $x^2 - dy^2 = n^2$ has solutions in addition to $x = n, y = 0$] was only filled by Lagrange. This, I believe, is one of the most important achievements of this great mathematician in the area of number theory because the tools he introduced for this purpose can be very well generalized and applied to analogous higher problems.

References

J. Itard: Lagrange, Joseph Louis (in: *Dictionary of Scientific Biography*).
I. Niven and H. S. Zuckermann: *An Introduction to the Theory of Numbers*, Wiley, New York, London, Sydney, 1966.

CHAPTER 5
Legendre

One of the most celebrated theorems in number theory is the law of quadratic reciprocity. We formulated it at the end of Chapter 3. The history of the discovery of this theorem is complicated and not quite clear, but we will shortly show that one is led to the theorem by the problem of deciding whether a given prime number divides a number of the form $x^2 + ay^2$. This was how Euler and later (around 1785), independently, Legendre discovered the theorem. Unlike Euler, Lagrange tried to prove the theorem, but his proof had serious gaps. We will discuss it below. Finally, it was rediscovered by Gauss, probably after numerical calculations and not in connection with the theory of binary forms. Gauss gave the first complete proof.

Let p be an odd prime number and a an integer with $(p, a) = 1$. Legendre defined the following symbol:

$$\left(\frac{a}{p}\right) := \begin{cases} 1, & \text{if the congruence } x^2 \equiv a \bmod p \text{ is solvable,} \\ -1, & \text{otherwise.} \end{cases}$$

Today, $\left(\frac{a}{p}\right)$ is called the Legendre Symbol. In the first case, a is called a quadratic residue modulo p, in the second, a quadratic nonresidue modulo p.

(5.1) **Theorem.** *Let p, q be prime numbers $\neq 2$. Then*:

$$\left(\frac{p}{q}\right)\left(\frac{q}{p}\right) = (-1)^{(1/4)(p-1)(q-1)}. \tag{1}$$

$$\left(\frac{-1}{p}\right) = \begin{cases} 1, & \text{if } p \equiv 1 \mod 4 \\ -1, & \text{if } p \equiv 3 \mod 4 \end{cases}$$

$$= (-1)^{(p-1)/2},$$

(2)

$$\left(\frac{2}{p}\right) = \begin{cases} 1, & \text{if } p \equiv 1,7 \mod 8 \\ -1, & \text{if } p \equiv 3,5 \mod 8 \end{cases}$$

$$= (-1)^{1/8(p^2-1)}.$$

(3)

Formula (1) is called the *law of quadratic reciprocity*.

(2) has already been proved [see (2.13)]. It is called the *first supplement to the law of quadratic reciprocity*.

(3) is called the *second supplementary theorem*. (1) establishes a connection between $(\frac{p}{q})$ and $(\frac{q}{p})$. Offhand, it is not immediately clear that these two expressions are in any way related.

We will come to the proof of the theorem later, but first we will discuss what it means.

If p is an odd prime number, the multiplicative group $\mathbb{F}_p^*$ of the field $\mathbb{F}_p$ with p elements is cyclic of order $p-1$. The kernel of the homomorphisms $\mathbb{F}_p^* \ni x \mapsto x^2 \in \mathbb{F}_p^*$ has order 2. Therefore, $(\mathbb{F}_p^*)^2$, the image of this homomorphism, has order $(p-1)/2$. This means that $\mathbb{F}_p^*$ contains the same number of squares as nonsquares: $[\mathbb{F}_p^* : (\mathbb{F}_p^*)^2] = 2$. Let $\bar{a}, \bar{b} \in \mathbb{F}_p^*$ be two nonsquares. Then the product $(\bar{a})(\bar{b})$ is a square. This leads to

$$\left(\frac{ab}{p}\right) = \left(\frac{a}{p}\right)\left(\frac{b}{p}\right).$$

In addition, trivially,

$$\left(\frac{a}{p}\right) = \left(\frac{a+kp}{p}\right).$$

For a "denominator" b that can be written $b = p_1 \ldots p_k$, one defines

$$\left(\frac{a}{b}\right) = \left(\frac{a}{p_1}\right) \cdots \left(\frac{a}{p_k}\right).$$

For odd a and b with $(a,b) = 1$ the following formula is a consequence of (5.1).

$$\left(\frac{a}{b}\right)\left(\frac{b}{a}\right) = (-1)^{(1/4)(a-1)(b-1)}.$$

Now we can easily compute the Legendre symbol. An example:

$$\left(\tfrac{383}{417}\right) = \left(\tfrac{417}{383}\right)(-1)^{(1/4)382\cdot416} = \left(\tfrac{34}{383}\right) = \left(\tfrac{17}{383}\right)\left(\tfrac{2}{383}\right)$$

$$= \left(\tfrac{17}{383}\right) \cdot 1 = \left(\tfrac{383}{17}\right)(-1)^{(1/4)382\cdot16} = \left(\tfrac{383}{17}\right) \cdot 1 = \left(\tfrac{9}{17}\right) = 1.$$

The following theorem establishes the connection between the representation of numbers by binary quadratic forms and quadratic reciprocity.

(5.2) **Theorem.** *Let m be a natural number properly represented by the form $ax^2 + 2bxy + cy^2$. Then $b^2 - ac$ is a quadratic residue modulo m.*

PROOF. Let x_0, y_0 be two relatively prime integers such that $m = ax_0^2 + 2bx_0y_0 + cy_0^2$. Let k, l be two integers with $kx_0 + ly_0 = 1$. Then

$$\left(ax_0^2 + 2bx_0y_0 + cy_0^2\right)(al^2 - 2bkl + ck^2)$$

$$= \left(k(x_0b + y_0c) - l(x_0a + y_0b)\right)^2 - (b^2 - ac)(kx_0 + ly_0)^2$$

or

$$m(al^2 - 2bkl + ck^2) = \left(k(x_0b + y_0c) - l(x_0a + y_0b)\right)^2 - (b^2 - ac).$$

Our statement follows.

Now we will show that the problem of representing a prime number by the form $x^2 + ay^2$ leads to the law of quadratic reciprocity, i.e., the connection between $(\frac{p}{q})$ and $(\frac{q}{p})$. If one can represent a prime number p of the form $4n + 3$ by $x^2 + ay^2$, then p, by (4.7), is not a divisor of $x^2 - ay^2$. This means that p cannot be represented by a form with discriminant $-a$. Hence $(\frac{a}{p}) = -1$. Conversely, if the condition $(\frac{a}{p}) = -1$ is satisfied for a prime number p, then, by (5.2), p cannot be represented by a form of discriminant $-a$. Then, by (4.7), p is a divisor of $x^2 + ay^2$. Let us now consider all reduced forms of discriminant a and look at congruences to determine whether or not p can be represented by $x^2 + ay^2$. We have already seen that this technique is often successful. The condition $(\frac{a}{p}) = -1$ is crucial for representing the prime number $p = 4n + 3$ by the form $x^2 + ay^2$.

To illustrate this situation, we investigate the representability by the form $x^2 + ay^2$ of prime numbers $p = 4n + 3$ of the special form $p = ka + b$. The number b is to be chosen in such a way that $(\frac{a}{ka+b}) = -1$. Offhand, it appears to be difficult to check this condition. More specifically, it is not at all clear that $(\frac{a}{ka+b})$ depends only on the residue of b. One could think that prime numbers of the form $ka + b$ might yield the symbol $+1$ or -1 and that sometimes we have a representation and sometimes not. Also, it would be very difficult to compute $(\frac{a}{p})$ for large $p = ka + b$. For the sake of simplicity, we assume that a is a prime number. Then it is easy to compute the "reciprocal" $(\frac{p}{a}) = (\frac{b}{a})$ which depends only on b. In other words, one studies whether it is helpful to know $(\frac{b}{a})$ and this leads to the law of quadratic reciprocity which then, in fact, solves the problem as follows.

By the law of quadratic reciprocity one has

$$\left(\frac{a}{p}\right) = \left(\frac{a}{ka+b}\right) = \left(\frac{ka+b}{a}\right)(-1)^{(1/4)(a-1)(p-1)}$$

$$= \left(\frac{b}{a}\right)(-1)^{(1/2)(a-1)(1/2)(p-1)} = \left(\frac{b}{a}\right)(-1)^{(1/2)(a-1)}.$$

Thus the condition $(\frac{a}{p}) = -1$ is satisfied if and only if $(\frac{b}{a})(-1)^{(1/2)(a-1)}$ $= -1$.

Now we give a few examples.

$a = 3$, $b = 1$. Then $(\frac{-1}{3})(-1)^{(3-1)/2} = -1$. We know that $p \equiv 1 \mod 3$, and consequently p is congruent 7 modulo 12. This means that every prime number $p \equiv 7$ modulo 12 is a divisor of $x^2 + 3y^2$; since the other reduced form with discriminant 3, $2x^2 + 2xy + 2y^2$, never has residue 7 modulo 12, every prime number of the form $p = 7$ modulo 12 can be represented by $x^2 + 3y^2$.

Let $a = 5$. Then $(\frac{b}{a})(-1)^{(a-1)/2} = (\frac{b}{5}) = -1$ if and only if $b = 2$ or 3. In these cases, p is congruent 3 or 7 modulo 20. Every prime number of the form $p = 3, 7$ modulo 20 is a divisor of $x^2 + 5y^2$ and thus can be represented by $x^2 + 5y^2$ or $2x^2 + 2xy + 3y^2$. But $x^2 + 5y^2 = x^2 + y^2$ modulo 4; consequently $x^2 + 5y^2$ represents at most the odd numbers congruent to 1 modulo 4. Consequently, prime numbers of the form $p \equiv 3, 7$ modulo 20 are represented by $2x^2 + 2xy + 3y^2$.

Let $a = 7$. Then $(\frac{b}{a})(-1)^{(a-1)/2} = (\frac{b}{7}) = -1$ if and only if $b = 1, 2$ or 4. p is congruent to $11, 15, 23$ modulo 28. By analogous reasoning, one sees that p can be represented by $x^2 + 7y^2$.

Like his slightly older contemporary Lagrange, Adrien Marie Legendre was the offspring of a wealthy family. He received a solid education at the Collège Mazarin in Paris and concluded his studies in mathematics and physics in 1770 when he was 18 years old. Abbé François-Joseph Marie, who had also furthered Lagrange's carrier, introduced Legendre to mathematics. Legendre was financially independent and was able to devote several years to pure research. Between 1775 and 1780, he was a teacher at the École Militaire in Paris. After 1783 he was connected with the Academie, first as successor of Laplace as "adjoint mécanicien" and then, from 1785 on, as "associé."

In 1782, Legendre won the prize of the Berlin Academie with a paper on ballistics. This is how he came to Lagrange's attention, who was still at Berlin. Later there were publications in number theory, celestial mechanics, and the theory of elliptic functions. During the French revolution, Legendre lost his fortune and was forced to give up his position at the Academie. The Commission for Public Affairs gave him the task of writing, together with Lagrange, a book on analysis and geometry. Legendre had several other public positions but he fell out of favor in 1824 and lost his annual pension of 3000 francs after he disagreed with certain official personnel policies. At his death in 1833, Legendre had a position at the

Bureau des Longitudes as successor of Lagrange. His priority quarrels with Gauss about the method of least squares and the law of quadratic reciprocity are dark chapters in Legendre's life. Quite upset and embittered, he complained to Jacobi that Gauss called both these discoveries his own. On the other hand, towards the end of his life, Legendre had the satisfaction of seeing that his favorite subject, the theory of elliptic functions, developed by two brilliant young mathematicians, Abel and Jacobi.

To this point, we have discussed the problem of the representation of natural numbers by binary quadratic forms. More generally, one can consider the analogous problem for forms in n variables for arbitrary $n \in \mathbb{N}$. We start out with a symmetric $n \times n$ matrix A with integral entries $a_{ij} = a_{ji}$, an n-tuple $x = (x_1, \ldots, x_n)$ of unknowns, and the quadratic form

$$q(x) = xAx^t = \sum_{i,j=1}^{n} a_{ij} x_i x_j$$
$$= \sum_{i=1}^{n} a_{ii} x_i^2 + \sum_{i<j} 2a_{ij} x_i x_j .$$

The general *representation problem* consists of finding necessary and sufficient conditions for the integral solutions and possibly also the number of solutions of the equation

$$q(x) = t$$

for a given $t \in \mathbb{Z}$. This very natural problem is extraordinarily difficult, and a complete solution is still far away.

Of course, a necessary condition for a solution of $q(x) = t$ is the solvability modulo an arbitrary prime power. (For reasons which we will not explain, one calls this "local" solvability of $q(x) = t$, whereas a solution of $q(x) = t$ is called "global.") One can easily see that the only important cases are powers of 2 and powers of those primes which are not relatively prime to the coefficients of the form. The example $5x^2 + 11y^2 = 1$ shows that, in general, this condition is not sufficient for global solvability. However, Legendre discovered an important case in which the "local-global principle" holds. He proved:

(5.3) **Theorem.** *Let a, b, c be integers other than 0, such that abc is square free. Then the equation*

$$ax^2 + by^2 + cz^2 = 0$$

has a nontrivial solution in integers if and only if a, b, c do not all have the same sign and $-bc, -ac, -ab$ are quadratic residue modulo a, b, and c, respectively. In other words, if the following congruences can be solved:

$$x^2 \equiv -bc \quad \mod a,$$
$$y^2 \equiv -ac \quad \mod b,$$
$$z^2 \equiv -ab \quad \mod c.$$

To see that this contains a "local-global principle" one first shows that the conditions on the signs of a, b, c can be replaced by the condition that the congruence

$$ax^2 + by^2 + cz^2 \equiv 0 \qquad \mod 8$$

is solvable in integers not all of which are even. To do this one has to distinguish several cases in a lengthy but simple proof. We will not go through this but show that the conditions thus modified for the solvability of $ax^2 + by^2 + cz^2 = 0$ are equivalent to the solvability of $ax^2 + by^2 + cz^2 = 0$ for every prime power N with g.c.d.$(x, y, z, N) = 1$. Obviously, the condition that the equation can be solved modulo every prime power is necessary for global solvability.

Conversely, assume that the equation $ax^2 + by^2 + cz^2 \equiv 0$ modulo N is solvable for every prime power N with g.c.d.$(x, y, z, N) = 1$. Specifically, let $N = p^2$ with $p \mid c$ and x_0, y_0, z_0 a solution of the corresponding congruence. Let y_0 be a multiple of p; because g.c.d.$(a, p) = 1$, x_0 is a multiple of p, contradicting our assumption g.c.d.$(x_0, y_0, z_0, N) = 1$. One similarly shows that x_0 is not a multiple of p, either. It follows from the congruence $ax_0^2 + by_0^2 \equiv 0$ modulo p that $-ab$ is quadratic residue modulo p and consequently modulo c. Analogously, $-ac$ is quadratic residue modulo p and $-bc$ a quadratic residue modulo a. One obtains the last condition for $N = 8$.

We will prove the above theorem of Legendre only in Chapter 9 because we will then have more efficient tools. Legendre tried to derive the law of quadratic reciprocity from (5.3). However, he was only partially successful. Nevertheless, we will follow some of his (basically) very interesting ideas, especially since they lead naturally to one of the most famous theorems in arithmetic.

Let us recall the main statement (1) of the law of quadratic reciprocity.

$$\left(\frac{p}{q}\right)\left(\frac{q}{p}\right) = (-1)^{(1/4)(p-1)(q-1)} = \begin{cases} -1, & \text{if } p, q \equiv 3 \mod 4, \\ 1, & \text{otherwise.} \end{cases} \tag{1}$$

The conditions $p, q \equiv 1, 3$ modulo 4 and $\left(\frac{p}{q}\right), \left(\frac{q}{p}\right) = \pm 1$ lead to 16 possibilities; to prove statement (1), one has to exclude half of these possibilities.

A first case can be excluded in the following way. For $p, q \equiv 3$ modulo 4, $\left(\frac{p}{q}\right) = 1$ and $\left(\frac{q}{p}\right) = 1$ can not hold simultaneously because then all the conditions of (5.3) for the solvability of the equation $ax^2 + by^2 + cz^2 = 0$ would be satisfied with $a = 1$, $b = -p$, $c = -q$. We know, however, that $x^2 - py^2 - qz^2 \equiv x^2 + y^2 + z^2$ modulo 4; this means, as one can easily see, that this equation has nontrivial solutions only for even x, y, z—a contradiction. Similarly, one excludes the case $p \equiv 1$ modulo 4, $q \equiv 3$ modulo 4, $\left(\frac{q}{p}\right) = 1$, $\left(\frac{p}{q}\right) = -1$ by looking at the equation $x^2 + py^2 - qz^2 = 0$. To exclude further cases, Legendre formulates and uses the following theorem which, however, he was not able to prove.

(5.4) **Theorem.** *Let m be a natural number and a an integer which is relatively prime to m. Then there are infinitely many prime numbers of the form $km + a$.*

This famous theorem was proved by Dirichlet (around 1837). We will discuss it in Chapter 8.

Using this theorem one can exclude further cases. Let us look at the case $p \equiv 1$ modulo 4, $q \equiv 3$ modulo 4, $(\frac{p}{q}) = 1$, $(\frac{q}{p}) = -1$. Then there is a prime number r with $r \equiv 1$ modulo 4 and $(\frac{r}{p}) = -1$, $(\frac{r}{q}) = -1$. To see this, we reason in the following way. The set of all numbers which are smaller than $4pq$ and relatively prime to $4pq$ consists of $2(p-1)(q-1)$ elements and decomposes into four classes. One of these classes consists of nonresidues of the numbers p and q. Half of this class consists of numbers $\equiv 1$ modulo 4 and the other half of numbers $\equiv 3$ modulo 4. Hence in this class there are $\frac{1}{4}(p-1)(q-1)$ nonresidues of p and of $q \equiv 1$ modulo 4; we call them $g, g', g'', \ldots$. The numbers $g + k \cdot 4pq$, $g' + k \cdot 4pq$, $g'' + k \cdot 4pq$, $\ldots$ $k \in \mathbb{Z}$, are nonresidues of $p, q \equiv 1$ modulo 4. By (5.4) there are infinitely many prime numbers of this form. By the second case above, $(\frac{q}{r}) = -1$ for such a prime number r. Furthermore, we know that $(\frac{p}{r}) = -1$ since $p \equiv 1$ modulo 4, $r \equiv 1$ modulo 4, $(\frac{p}{r}) = 1$, $(\frac{r}{p}) = -1$ is impossible because otherwise one could again, using (5.4), find a prime number r', $r' \equiv 3$ modulo 4 with $(\frac{r'}{p}) = -1$, i.e., $(\frac{r'}{p}) = -1$ and $(\frac{rr'}{p}) = +1$. Also, then $(\frac{-p}{rr'}) = 1$, $(\frac{-p}{r'}) = 1$, and consequently $(\frac{-p}{rr'}) = 1$. Then $x^2 + py^2 - rr'z^2 = 0$ would have nontrivial solutions. Calculating modulo 4, one sees that this is clearly impossible. On the other hand, from $(\frac{q}{r}) = -1$ and $(\frac{p}{r}) = -1$ one obtains $(\frac{pq}{r}) = 1$. Consequently, if we had $(\frac{p}{q}) = 1$, $(\frac{q}{p}) = -1$, then all the conditions for the existence of nontrivial solutions of $px^2 - qy^2 + rz^2 = 0$ would be satisfied. Again reducing modulo 4, one sees that this is a contradiction.

References

Y. Itard: Legendre, Adrien Marie (in *Dictionary of Scientific Biography*).

A. M. Legendre: *Theorie des Nombres*, Paris, 1830; reprint, Blanchard, Paris 1955.

A. M. Legendre: *Zahlentheorie*. (According to the Third Edition translated into German by H. Maser, Vol. II, Teubner, Leipzig, 1886.)

C. F. Gauss: *Disquisitiones Arithmeticae*, Art. 151 and 296.

Legendres correspondence with Jacobi in Jacobi, *Werke*, Vol. 1.

L. Kronecker: see references to Chapter 3.

CHAPTER 6
Gauss

Carl Friedrich Gauss lived from 1777 to 1855. In his lifetime he was known as "princeps mathematicorum." His main number-theoretical work, *Disquisitiones Arithmeticae*, and several smaller number-theoretical papers contain so many deep and technical results that we have to confine ourselves to just a small sample. Other equally important results will not be mentioned.

Gauss's mathematical career started in a singularly spectacular way. We know this from his diary which informs us about his most important discoveries. It begins with an entry on March 30, 1796: "Principia quibus innititur secticio circuli, ac divisibilitas eiusdem geometrica in septemdecim partes etc." In his letter of January 6, 1819 to Gerling, Gauss gives a more extensive description of his discovery of the constructability of the regular 17-gon:

> By concentrated analysis of the connection of all roots (of the equation $1 + x + \cdots + x^{p-1} = 0$) according to arithmetical reasons I succeeded, during a vacation in Braunschweig, in the morning of this day, before I got up, to see the connection clearly such that I was able to make the specific application to the 17-gon and to confirm it numerically right away.

In other words, Gauss solved an ancient problem:

(6.1) **Theorem.** *The regular 17-gon can be constructed with ruler and compass.*

As he emphasized in his very first announcement, Gauss's methods suffice to solve the problem of constructing regular *n*-gons completely.

(6.2) **Theorem.** *The regular n-gon can be constructed with ruler and compass if and only if $n = 2^k p_1 \ldots p_r$, where p_i are Fermat prime numbers, i.e., of the form $2^{2^i} + 1$.*

A few weeks later, on April 18, 1796, Gauss found the first complete proof of the law of quadratic reciprocity. He had discovered the law a few months earlier, independent of Euler and Legendre. In connection with this law he developed the theory of binary quadratic forms and established a theory which far transcends everything that had been done by his predecessors, specifically Lagrange and Legendre. His famous *Disquisitiones Arithmeticae*, first published in 1801 (in Latin), deals mainly with this subject; it established number theory as a systematic, well-founded, and rich area of mathematics.

Today, the theorem about constructing the regular *n*-gon is a part of Galois theory; but there are, as Gauss himself emphasizes, essential number-theoretical aspects to it. We will now sketch one of these, particularly since we will see that there is a connection to the law of quadratic reciprocity. Again, we see a process quite typical in the development of a mathematical theory: initially, one has a number of apparently unconnected facts, here the construction of the *n*-gon and the law of quadratic reciprocity. After penetrating into these problems more deeply, one discovers a close connection, often to the surprise of the discoverer; this prepares the way for what Eudemus called "reason."

The edges of the *n*-gon inscribed in the unit circle are the *n*th roots of unity,

$$\exp(2k\pi i/n) = \cos(2k\pi/n) + i\sin(2k\pi/n), \qquad k = 0, \ldots, n - 1.$$

These are the roots of the equation

$$x^n - 1 = (x - 1)(x^{n-1} + x^{n-2} + \cdots + 1) = 0.$$

Elementary algebra shows that the construction can be performed by ruler and compass if the equation

$$x^{n-1} + x^{n-2} + \cdots + 1 = 0$$

can be reduced to a chain of quadratic equations. For any prime number $n = p$, this can be done if and only if $p - 1$ is a power of 2 (as one knows from the Galois theory of the fields of roots of unity). Gauss determines the first one in this chain of quadratic equations in the following way. Let $\epsilon := \exp(2\pi i/p)$; the roots of $x^p - 1$ are $1, \epsilon, \ldots, \epsilon^{p-1}$; specifically, $1 + \epsilon + \cdots + \epsilon^{p-1} = 0$. Now consider the (as we say today) Gaussian sum

$$S := \sum_{k=1}^{p-1} \left(\frac{k}{p}\right)\epsilon^k,$$

where $(\frac{k}{p})$ is the Legendre symbol. Then:

(6.3) **Remark.** $S^2 = (\frac{-1}{p})p$. Thus the first field in the chain of quadratic field extensions $\mathbb{Q} \subset K_1 \subset K_2 \subset \cdots \subset K_n = \mathbb{Q}(\epsilon)$ is $\mathbb{Q}(\sqrt{(\frac{-1}{p})p}$.

PROOF. We know that

$$S^2 = \sum_{k,l=1}^{p-1} \left(\frac{k}{p} \right)\left(\frac{l}{p} \right)\epsilon^{k+l}$$

$$= \sum_{k,l} \left(\frac{kl}{p} \right)\epsilon^{k+l}.$$

As k runs through the nonzero residue classes modulo p, kl will as well, with l fixed. We may therefore replace k by kl:

$$S^2 = \sum_{k,l} \left(\frac{kl^2}{p} \right)\epsilon^{kl+l}$$

$$= \sum_{k,l} \left(\frac{k}{p} \right)\epsilon^{l(k+1)}$$

$$= \sum_{l} \left(\frac{-1}{p} \right)\epsilon^0 + \sum_{k \neq p-1} \left(\frac{k}{p} \right)\left(\sum_{l} \epsilon^{l(k+1)} \right),$$

and, because $\sum_{l} \epsilon^{l(k+1)} = \epsilon + \epsilon^2 + \cdots + \epsilon^{p-1} = -1$, this expression can be written as

$$= \left(\frac{-1}{p} \right)(p-1) + \sum_{k \neq p-1} \left(\frac{k}{p} \right) \cdot (-1)$$

$$= \left(\frac{-1}{p} \right)(p-1) - \left(\frac{-1}{p} \right)(-1)$$

$$= \left(\frac{-1}{p} \right)p.$$

In other words, (6.3) determines the Gaussian sum S up to sign. An obvious and, as we shall see, for many problems, very important question is to determine the sign of S. However, this is a very difficult problem and Gauss worked very hard, over several years, before finding the solution. He writes in a letter to Olbers (September 3, 1805):

> The determination of the sign of the root has vexed me for many years. This deficiency overshadowed everything that I found; over the last four years, there was rarely a week that I did not make one or another attempt, unsuccessfully, to untie the knot. Finally, a few days ago, I succeeded—but not as a result of my search but rather, I should say, through the mercy of God. As lightning strikes, the riddle has solved itself.

(6.4) **Theorem** (Gauss, Summatio quarundam serierum singularium, 1808, *Werke* II). *Let* $\epsilon := \exp(2\pi i/p)$. *Then*

$$S = \sum_{k=1}^{p-1}\left(\frac{k}{p}\right)\epsilon^k = \begin{cases} \sqrt{p}, & \text{if } p \equiv 1 \mod 4, \\ i\sqrt{p}, & \text{if } p \equiv 3 \mod 4 \end{cases}$$

(*choose the positive square root*).

The proof of this theorem is an easy consequence of the following theorem which has a slightly different Gaussian sum as its subject. Let m be a natural number and

$$G(m) := \sum_{k=0}^{m-1} \epsilon^{k^2} \quad \text{with} \quad \epsilon := \exp\left(\frac{2\pi i}{m}\right).$$

(6.5) **Theorem.** *We have the following*:

$$G(m) = \begin{cases} (1+i)\sqrt{m} & \text{for } m \equiv 0 \mod 4, \\ \sqrt{m} & \text{for } m \equiv 1 \mod 4, \\ 0 & \text{for } m \equiv 2 \mod 4, \\ i\sqrt{m} & \text{for } m \equiv 3 \mod 4. \end{cases}$$

Later we will give Dirichlet's beautiful proof of this theorem. Here we derive (6.4) from (6.5) (this is a simple exercise). Then, following Gauss, we prove the law of quadratic reciprocity.

Derivation of (6.4) from (6.5). Let μ run through all the quadratic residues and ν through all the nonresidues modulo p. Then obviously

$$S = \sum_{\mu}\epsilon^{\mu} - \sum_{\nu}\epsilon^{\nu}.$$

Since

$$1 + \sum_{\mu}\epsilon^{\mu} + \sum_{\nu}\epsilon^{\nu} = 0,$$

we have

$$S = 1 + 2\sum_{\mu}\epsilon^{\mu}.$$

Let us assume k runs through all the numbers $0, 1, \ldots, p-1$. Then k^2 runs through all the quadratic residues exactly twice, except for 0 which occurs only once. Consequently,

$$S = \sum_{k=0}^{p-1} \epsilon^{k^2} = G(p),$$

i.e., (6.4).

We now prove the formula

$$\left(\frac{p}{q}\right)\left(\frac{q}{p}\right) = (-1)^{(1/4)(p-1)(q-1)}$$

for odd primes p, q. If $k \equiv 1$ modulo m, then $(k^2 - l^2)/m \in \mathbb{Z}$, i.e., $\exp(2\pi i k^2/m) = \exp(2\pi i l^2/m)$. This means that we can compute the Gaussian sum using an arbitrary system of residues modulo m. Specifically,

$$G(2m) = \sum_{k=-m}^{m-1} \epsilon^{k^2} = 2 \sum_{k=0}^{m-1} \epsilon^{k^2} - 1 + \epsilon^{m^2}$$

$$= 2 \sum_{k=0}^{m-1} \epsilon^{k^2} - 1 + (-1)^m.$$

By (6.5), one has for even $m = 2n$

$$\sum_{k=0}^{2n-1} \exp\left(\frac{2\pi i k^2}{4n}\right) = \frac{1}{2} G(4n) = (1 + i)\sqrt{n} .$$

Let

$$H(2n) := \sum_{k=0}^{2n-1} \exp\left(\frac{2\pi i k^2}{4n}\right).$$

Also, here, we can compute the sum with the help of an arbitrary residue system for $(k + 2nl)^2 = k^2 + 4nkl + 4n^2l^2 \equiv k^2$ modulo $4n$. Now let p, q be two relatively prime odd numbers. We claim that the following formula holds:

$$H(4pq) = \left(\sum_{\mu=1}^{4} \exp\left(\frac{2\pi i \mu^2 pq}{8}\right)\right)\left(\sum_{\nu=1}^{p} \exp\left(\frac{2\pi i 2 q \nu^2}{p}\right)\right)\left(\sum_{\rho=1}^{q} \exp\left(\frac{2\pi i 2 p \rho^2}{q}\right)\right).$$

One sees this by observing that $k = \mu pq + \nu 4q + \rho 4p$ goes through a full system of residues modulo $4pq$ for $1 \leq \mu \leq 4$, $1 \leq \nu \leq p$, $1 \leq \rho \leq q$. Then

$$\exp\left(\frac{2\pi i k^2}{8pq}\right) = \exp\left(\frac{2\pi i(\mu^2 p^2 q^2 + \nu^2 16 q^2 + \rho^2 16 p^2)}{8pq}\right).$$

Since $\exp(2\pi i t)$ has period 1, the mixed terms of k^2 can be deleted. This immediately yields our formula. For the sake of brevity, we write

$$H(4pq) = H_2(4pq)H_p(4pq)H_q(4pq)$$

and compute the three factors separately. With $\eta = \exp(2\pi i/8)$ one obtains

$$H_2 = \eta^{pq} + \eta^{4pq} + \eta^{9pq} + \eta^{16pq} = 2\eta^{pq}$$

because $\eta^8 = 1$. First we calculate H_p when $q = 1$. Then

$$(1 + i)\sqrt{2p} = H(4p) = H_2H_p = 2\eta^p H_p,$$

and consequently

$$H_p(4p) = H_p = \eta^{1-p}\sqrt{p} \qquad \left(\eta = \frac{1+i}{\sqrt{2}}\right).$$

We now need a lemma.

Lemma. *Let p, q be different odd primes. Then*

$$H_p(4pq) = \sum_{\nu=1}^{p} \exp\left(\frac{2\pi i 2 q \nu^2}{p}\right) = \left(\frac{q}{p}\right)\sqrt{p}\,\eta^{1-p}.$$

PROOF. Let $\left(\frac{q}{p}\right) = 1$. Then $2\nu^2$ runs through the same numbers as $2q\nu^2$ modulo p. Thus our claim follows from the case $q = 1$ considered above. If $\left(\frac{q}{p}\right) = -1$, then $q\nu^2$ runs through the quadratic nonresidues modulo p twice except for 0 which occurs only once. Then

$$\sum_{\nu=1}^{p} \exp\left(\frac{2\nu^2}{p}\right) + \sum_{\nu=1}^{p} \exp\left(\frac{2q\nu^2}{p}\right) = 0,$$

for the complete sum is twice the sum of all the pth roots of unity. This equation together with the first case completes the proof.

We are now nearly finished with our proof of the law of quadratic reciprocity. From what we have shown so far, it follows that

$$2\sqrt{pq}\,\eta = H(4pq) = H_2H_pH_q$$

$$= 2\eta^{pq}\left(\frac{q}{p}\right)\sqrt{p}\,\eta^{1-p}\left(\frac{p}{q}\right)\sqrt{q}\,\eta^{1-q}$$

and consequently

$$\left(\frac{p}{q}\right)\left(\frac{q}{p}\right) = \eta^{(p-1)(q-1)} = (-1)^{(1/4)(p-1)(q-1)}.$$

This is Gauss's fourth proof of the law of quadratic reciprocity. In 1818 he published another proof (his sixth) which also uses the theory of Gauss sums but needs only the simple result (6.3):

$$S^2 = \left(\frac{-1}{p}\right)p,$$

and the congruence

$$(x + y)^q \equiv x^q + y^q \pmod{q}$$

for any prime number q. Then one has

$$\left(\sum_{k=1}^{p-1}\left(\frac{k}{p}\right)\epsilon^k\right)^q \equiv \sum_{k=1}^{p-1}\left(\frac{k}{p}\right)\epsilon^{qk} \pmod{q},$$

and thus

$$S^q \equiv \left(\frac{q}{p} \right) S \pmod{q}.$$

We now multiply by S and use (6.3):

$$\left(\frac{-1}{p} \right)^{(q+1)/2} p^{(q+1)/2} \equiv \left(\frac{q}{p} \right) \left(\frac{-1}{p} \right) p \pmod{q}$$

$$\left(\frac{-1}{p} \right)^{(q-1)/2} p^{(q-1)/2} \equiv \left(\frac{q}{p} \right) \pmod{q}.$$

For every number a relatively prime to q, the so-called Euler criterion holds:

$$a^{(q-1)/2} \equiv \left(\frac{a}{q} \right) \mod q$$

(which is proved by using the fact that the residues form a cyclic group). Then it follows that

for $p \equiv 1 \mod 4$: $\qquad \left(\dfrac{p}{q} \right) = \left(\dfrac{q}{p} \right);$

for $p \equiv 3 \mod 4$: $\qquad \left(\dfrac{-1}{q} \right) \left(\dfrac{p}{q} \right) = \left(\dfrac{q}{p} \right).$

These formulas contain the quadratic law of reciprocity. We cheated a bit in our proof because right at the beginning we used the congruence $(x + y)^q \equiv (x^q + y^q) \mod q$ for nonintegral x, y. One justifies this step with a bit of algebra. Instead of making our calculations in $\mathbb{Z}$, we make them in the ring

$$\mathbb{Z}[\epsilon] := \mathbb{Z} \oplus \mathbb{Z}\epsilon \oplus \cdots \oplus \mathbb{Z}\epsilon^{p-2}, \qquad \epsilon = \exp\left(\frac{2i}{p} \right).$$

Virtually the same proof was published later by Jacobi, Eisenstein, and Cauchy, who involved themselves in a priority quarrel over it.

We can prove the second supplement of the law of quadratic reciprocity with the help of Euler's criterion. Since

$$\frac{(1 + i)^2}{i} = 2$$

Euler's criterion yields

$$\left(\frac{2}{p} \right) \equiv (1 + i)^{p-1} / i^{(p-1)/2} \mod p$$

$$\equiv (1 + i)^p / i^{(p-1)/2}(i + 1) \mod p$$

$$\equiv (1 + i^p) / i^{(p-1)/2}(1 + i) \mod p$$

$$\equiv \frac{\exp(p\pi i/4) + \exp(-p\pi i/4)}{\exp(\pi i/4) + \exp(-\pi i/4)} \mod p$$

$$\equiv \frac{\cos(p\pi/4)}{\cos(\pi/4)} \mod p.$$

Then

$$\left(\frac{2}{p}\right) = \frac{\cos(p\pi/4)}{\cos(\pi/4)} = \begin{cases} 1 & \text{if } p \equiv \pm 1 \mod 8, \\ -1 & \text{if } p \equiv \pm 3 \mod 8, \end{cases}$$

and consequently

$$\left(\frac{2}{p}\right) = (-1)^{(p^2-1)/8}.$$

Gauss, who called it his *theorema fundamentale*, considered the law of quadratic reciprocity to be one of his most important contributions to number theory. Again and again he came back to it and altogether proved it in six different ways. His first proof, which uses only properties of the integers, is the most elementary. In a simplified and elegant form it was presented by Dirichlet (see his *Werke II*, page 121). With this we conclude our discussion of Gaussian sums and the law of quadratic reciprocity.

Up until now, all our manipulations were possible without leaving the real numbers. Since Cardano's time, mathematicians were dimly aware of the existence of the complex numbers, but when Gauss started his work it was not yet obvious how to use them. Their naive manipulation, e.g., in Euler's work, led to a number of mistakes. Gauss's dissertation dealt with the fundamental theorem of algebra (for which d'Alembert had already given a nearly complete proof). Interest in the fundamental theorem of algebra was motivated by the decomposition of rational functions into partial fractions, the decomposition being needed for integrating these functions.

(6.6) **Theorem** (Fundamental Theorem of Algebra). *Every polynomial with complex coefficients can be written as a product of linear factors over the field of complex numbers.*

For number theory it was more important that Gauss considered also *integral* complex numbers. This initiated a development which was taken up and pursued by the most important number theorists of the nineteenth century. Gauss's most significant achievement in this field might well be the discovery and the proofs of the cubic and biquadratic law of reciprocity. These two laws deal with the behavior of integers modulo 3rd and 4th powers. We will not discuss this question in any detail but will instead show that the integral complex numbers provide a conceptual framework for simple and elegant solutions of many number-theoretical problems. Most of the problems we will discuss had already been posed by Fermat. For the sake of simplicity, we will use modern terminology.

As usual, let $i = \sqrt{-1}$ and $A := \{a + bi \mid a, b \in \mathbb{Z}\}$. One can immediately see that A is closed with respect to addition and multiplication. We call A the ring of Gaussian integers. As is well known, the elements of A correspond to the points of a lattice in the complex plane. The complex number $x = a + bi$ has the "norm"

$$\|x\| = a^2 + b^2 = (a + bi)(a - bi).$$

(This already indicates that there is a connection to the representation of numbers as sums of two squares.) According to (2.1) we have $\|xy\| = \|x\|\|y\|$. The arithmetic of A is based on the following simple theorem.

(6.7) Theorem (Gauss). *A is a Euclidean ring, i.e., for every $x, y \in A$, $y \neq 0$, there are $q, r \in A$ with*

$$x = qy + r, \qquad r = 0 \quad or \quad \|r\| < \|y\|.$$

PROOF. Since $y \neq 0$ we may form the complex number

$$\frac{x}{y} = \alpha + \beta i; \qquad \alpha, \beta \in \mathbb{Q}.$$

Then $a, b \in \mathbb{Z}$ exist with $|a - \alpha|, |b - \beta| \leqslant \frac{1}{2}$. For $q := a + bi$ it follows that $\|x/y - q\| \leqslant \frac{2}{4}$. Also, $x = qy + (x - qy)$ with

$$x - qy = 0 \quad or \quad \|x - qy\| \leqslant \frac{2}{4}\|y\| < \|y\|.$$

A well-known consequence of the existence of a Euclidean algorithm is that every element of A can be written uniquely as a product of prime elements. Speaking more precisely, this representation is unique up to multiplication by units, i.e., invertible elements, of the ring A. If x is a unit element and $xy = 1$ it follows that $\|x\|\|y\| = 1$, and consequently $\|x\| = 1$, i.e., $x \in \{1, -1, i, -i\}$. If π is a prime element, $\pi, -\pi, i\pi, -i\pi$ are "associated" prime elements, i.e., they differ from π only by a unit. If one selects one of these four elements, every element of A can uniquely, up to order, be written in the form

$$\epsilon \pi_1 \dots \pi_n; \qquad \epsilon \text{ unit}, \quad \pi_i \text{ prime element}. \tag{6.8}$$

Let us mention that Gauss obviously realized that it is necessary to prove unique decomposition into prime factors in A. Other mathematicians, among them Euler, along with his predecessors and several mathematicians of the following generations (Lamé, Cauchy, Kummer), implicitly use this property (in other rings) without proving it or, occasionally, even when it does not hold. (In this connection see remarks in Edwards, *Fermat's Last Theorem*, page 76ff).

If one wants to determine all the prime elements of A, it suffices to determine the decomposition of the ordinary (rational) primes into prime factors. This is closely related to the representation of numbers as sums of two squares.

(6.9) **Theorem** (Fermat–Euler). *The prime numbers can be written as products of prime elements in A in the following way*:

$$2 = (-i)(1 + i)^2, \qquad 1 + i \; a \; prime \; element,$$

$$p = p, \qquad\qquad\qquad p \; a \; prime \; element, \; when \; p \equiv 3 \bmod 4,$$

$$p = q_1 q_2, \qquad\qquad q_1, q_2 \; nonassociated \; prime \; elements, \; if \; p \equiv 1 \bmod 4.$$

PROOF. Let $a + bi$ be a prime element, then $a - bi$ is a prime element as well (since, if it were possible to decompose $a - bi$ into two factors, one could do this for $a + bi$ analogously). Then $a^2 + b^2$ has the following decomposition into prime factors:

$$a^2 + b^2 = (a + bi)(a - bi).$$

Consequently, $a^2 + b^2 = p$ or $a^2 + b^2 = p^2$, where p is a rational prime. If $p \equiv 3 \bmod 4$, then $a^2 + b^2 = p$ cannot be solved, i.e., p is a prime element of A. If $p \equiv 1 \bmod 4$, then, by (2.3),

$$a^2 + b^2 = p$$

is solvable, and we can write

$$p = q_1 q_2, \qquad q_1 = (a + bi), \qquad q_1 = (a - bi).$$

The statement is trivial for $p = 2$.

Remark. Here we see once more that a prime number $p \equiv 1 \bmod 4$ can, up to sign and order of the summands, be represented uniquely as a sum of two squares; if this were not the case, p could be written in two different ways as a product of prime elements in A, contradicting (6.8).

Having introduced the ring of Gaussian integers, we will now indicate how number theory can be further developed in this ring. To do this, we define a ζ-function for A:

$$\zeta_A(s) := \sum_{\substack{a \in A \\ a \neq 0}} \frac{1}{\|a\|^s} \, .$$

Since every element $a \in A$ can be written uniquely as a product of prime elements (and units) this series can also be expanded as an Euler product:

$$\zeta_A(s) = 4 \prod_q \left(1 + \frac{1}{\|q\|^s} + \frac{1}{\|q\|^{2s}} + \cdots \right) = 4 \prod_q \left(\frac{1}{1 - \|q\|^{-s}} \right);$$

$\{q\}$ is a system of representatives for the prime elements of A. The factor 4 is due to the four units. The prime elements were listed in (6.9). For the last product this yields

$$\zeta_A(s) = 4\left(\frac{1}{1-2^{-s}}\right)\prod_{p\equiv1(4)}\left(\frac{1}{1-p^{-s}}\right)^2\prod_{p\equiv3(4)}\left(\frac{1}{1-p^{-2s}}\right).$$

The first factor stems from the prime element $1 + i$ with $\|1 + i\| = 2$. Next is the product coming from the two factors of p with norm p, and the last product comes from the prime elements of norm p^2. This leads to

$$\zeta_A(s) = 4\left(\frac{1}{1-2^{-s}}\right)\prod_{p\equiv1(4)}\left(\frac{1}{1-p^{-s}}\right)^2\prod_{p\equiv3(4)}\left(\frac{1}{1-p^{-s}}\right)\left(\frac{1}{1+p^{-s}}\right)$$

$$= 4\cdot\zeta(s)\cdot\prod_{p\equiv1(4)}\left(\frac{1}{1-p^{-s}}\right)\prod_{p\equiv3(4)}\left(\frac{1}{1+p^{-s}}\right)$$

$$= 4\cdot\zeta(s)\cdot L(s).$$

In this formula, $\zeta(s)$ is the zeta-function which we introduced in Chapter 3, and $L(s)$ is a so-called L-function,

$$L(s) := \prod_{p\equiv1(4)}\left(\frac{1}{1-p^{-s}}\right)\prod_{p\equiv3(4)}\left(\frac{1}{1+p^{-s}}\right)$$

$$= \prod_{p\equiv1(4)}\left(1+\frac{1}{p^s}+\frac{1}{p^{2s}}+\cdots\right)\prod_{p\equiv3(4)}\left(1-\frac{1}{p^s}+\frac{1}{p^{2s}}-\cdots\right).$$

Performing the multiplication yields, similar to the case of the ζ-function

$$L(s) = \sum_{n=1}^{\infty}\frac{\chi(n)}{n^s} = 1 - \frac{1}{3^s} + \frac{1}{5^s} - \frac{1}{7^s} + \frac{1}{9^s} - + \cdots$$

with

$$\chi(n) = \begin{cases} 0, & \text{if } n \text{ even,} \\ 1, & \text{if } n \equiv 1 \quad \mod 4, \\ -1, & \text{if } n \equiv -1 \quad \mod 4. \end{cases}$$

It is easy to see that $L(s)$ converges for $s > 0$. When $s = 1$ we have exactly the Leibniz series

$$L(1) = 1 - \tfrac{1}{3} + \tfrac{1}{5} - \tfrac{1}{7} + \tfrac{1}{9} - + \cdots$$

which we already know. Let us once more consider the equation

$$\zeta_A(s) = 4\zeta(s)L(s).$$

It has a pole of order 1 at $s = 1$ (see (3.9)). $\lim_{s\downarrow1}(s - 1)\zeta(s) = 1$ leads to the equation

$$\lim_{s\downarrow1}(s - 1)\zeta_A(s) = 4L(1).$$

Since $\|x + iy\| = x^2 + y^2$,

$$(s - 1)\zeta_A(s) = (s - 1)\sum_{(x,y)\neq(0,0)}\frac{1}{\left(x^2 + y^2\right)^s}.$$

Intuitively, it is fairly clear that we have an approximate equality of the following kind (the left-hand side is a Riemann sum for the integral):

$$\sum_{(x,y)\neq(0,0)} \frac{1}{\left(x^2+y^2\right)^s} \approx \iint_{x^2+y^2\geq 1} \frac{dx\,dy}{\left(x^2+y^2\right)^s}.$$

More precisely, one has

$$\lim_{s\downarrow 1}(s-1)\left(\sum_{(x,y)\neq(0,0)} \frac{1}{\left(x^2+y^2\right)^s} - \iint_{x^2+y^2\geq 1} \frac{dx\,dy}{\left(x^2+y^1\right)^s}\right) = 0.$$

Computing this integral is a standard exercise in analysis. One substitutes polar coordinates $x = r\cos\phi$, $y = r\sin\phi$ and obtains

$$\iint_{x^2+y^2\geq 1} \frac{dx\,dy}{\left(x^2+y^2\right)^s} = \int_0^{2\pi}\int_1^\infty \frac{r\,dr\,d\phi}{r^{2s}} = 2\pi\left[\frac{1}{-2(s-1)r^{2(s-1)}}\right]_1^\infty$$

$$= \frac{\pi}{s-1}.$$

This yields

$$\lim_{s\downarrow 1}(s-1)\zeta_A(s) = \lim_{s\downarrow 1}(s-1)\iint_{x^2+y^2\geq 1} \frac{dx\,dy}{\left(x^2+y^2\right)^s} = \pi$$

and finally the Leibniz series

$$\frac{\pi}{4} = L(1) = 1 - \frac{1}{3} + \frac{1}{5} - \frac{1}{7} + \frac{1}{9} - + \cdots.$$

We have thus proved Leibniz' formula once more. Since this proof is much more complicated than the usual proof in Chapter 3, one is led to ask about the purpose of our calculation. To answer this, let us remember what is at the root of our proof, namely the unique decomposition into prime factors in A. Offhand, it is rather surprising that this and the Leibniz series for $\pi/4$ should be connected, but this connection is in fact of utmost importance in number theory and was discovered in full generality by Dirichlet. Gauss, whose proof of Leibniz' formula is quite similar, never published his ideas, but some of them, only partially developed, are in a paper that was written 33 years after *Disquisitiones Arithmeticae* and published posthumously. It would be interesting to know whether he gave Dirichlet any clues in this regard. Unfortunately, we cannot tell; Dirichlet's papers that deal with this subject do not refer to Gauss; conversely, Gauss did not make any comments on those of Dirichlet's papers that develop these ideas. In general, Gauss paid very little attention to the discoveries of other mathematicians. This is particularly regrettable in Dirichlet's case because he was one of the few from whom Gauss could have learned things which he did not already know.

Having strayed from our consideration of Gauss's main number-theoretical papers, we might as well stray a bit further. Let us consider once more the Zeta-function $\zeta_A(s)$. From its definition, and since $\|a\| = x^2 + y^2$, we immediately have

$$\zeta_A(s) = \sum_{n=1}^{\infty} \frac{N_2(n)}{n^s} ,$$

where $N_2(n)$ denotes the number of representations of n as the sum of two squares $x^2 + y^2$, $x, y \in \mathbb{Z}$. Considering the identity

$$\zeta_A(s) = 4\zeta(s)L(s)$$

$$= 4\left(\sum_{k=1}^{\infty} \frac{1}{k^s} \right)\left(\sum_{m=1}^{\infty} \frac{\chi(m)}{m^s} \right)$$

$$= 4 \sum_{n=1}^{\infty} \left(\sum_{m/n} \chi(m) \right) n^{-s},$$

one derives, by comparing coefficients, the following theorem. (Comparing coefficients is permitted because of a uniqueness theorem, similar to the one for power series; see Chapter 8).

(6.10) Theorem. $N_2(n) = 4 \sum_{m/n} \chi(m)$.

This theorem again solves one of Fermat's problems. The proof we have given was, we think, first found by Jacobi. We have already mentioned that Gauss did not work out his investigations of the Zeta-function ζ_A; but his posthumously published papers show that he knew much more general results. These will be discussed in Chapter 8.

More than four-fifths of the *Disquisitiones Arithmeticae* deals with quadratic congruences and binary quadratic forms. We know today that this theory is essentially equivalent to the ideal theory of quadratic numbers fields, i.e., extensions of the rational number field $\mathbb{Q}$ of degree 2. Gauss and his immediate successors, Dirichlet and Kummer, did not know this; only a generation later, with Dedekind, did this idea gain general acceptance. The language of ideals is much simpler, more perspicuous, and more suitable for generalizations and will be explained below. Our presentation goes back to Gauss in its substance, but the concepts and the way we present them are those of later mathematicians. If the reader is interested in the original version, he should consult the *Disquisitiones* or Dirichlet's *Number Theory*.

Let d be a square free integer $d \neq 1$. We consider the quadratic number field

$$\mathbb{Q}(\sqrt{d}) = \{ x + y\sqrt{d} \mid x, y \in \mathbb{Q} \}.$$

If $d > 0$, then $\mathbb{Q}(\sqrt{d}) \subset \mathbb{R}$, and the field is called a *real quadratic field*. If

$d < 0$ then $\mathbb{C} \supset \mathbb{Q}(\sqrt{d}) \not\subset \mathbb{R}$, and the field is an *imaginary quadratic field*. We define integers in the field $\mathbb{Q}(\sqrt{d})$ in the following way.

$$x + y\sqrt{d} \text{ integral} \Leftrightarrow \begin{cases} x, y \in \mathbb{Z} & \text{if } d \equiv 2, 3 \mod 4, \\ 2x, 2y, x + y \in \mathbb{Z} & \text{if } d \equiv 1 \mod 4. \end{cases}$$

(This definition is surprising when $d \equiv 1$ (mod 4); Gauss, Dirichlet, and Kummer did not think of this.) Let A_d be the set of integers in $\mathbb{Q}(\sqrt{d})$. For $\omega = \sqrt{d}$ for $d \equiv 2, 3$ and $\omega = \frac{1}{2}(1 + \sqrt{d})$ for $d \equiv 1$ we have

$$A_d = \{x + y\omega \mid x, y \in \mathbb{Z}\}.$$

Then, one easily sees:

Remark. A_d is a ring.

A_d is called the *ring of integers* in the quadratic number field $\mathbb{Q}(\sqrt{d})$. This ring can be characterized in the following way.

(6.11) **Remark.** $A_d = \{a \in \mathbb{Q}(\sqrt{d}) \mid a + a', aa' \in \mathbb{Z}\}$ (a' is the conjugate of a).

PROOF. Let $a = x + y\omega \in A_d$. It is straightforward to show that $a + a' \in \mathbb{Z}$. When $d \equiv 2, 3 \mod 4$, one has $(x + y\sqrt{d})(x - y\sqrt{d}) = x^2 - dy^2 \in \mathbb{Z}$; in the case $d \equiv 1 \mod 4$, $d = 1 + 4n$ one has

$$\left(x + y\frac{1 + \sqrt{d}}{2}\right)\left(x + y\frac{1 - \sqrt{d}}{2}\right) = x^2 + xy + \frac{y^2}{4} - \frac{y^2 d}{4}$$

$$= x^2 + xy + \frac{y^2}{4} - \frac{y^2}{4} - \frac{4ny^2}{4}$$

$$= x^2 + xy - ny^2 \in \mathbb{Z}.$$

Conversely, let $a = x + y\sqrt{d} \in \mathbb{Q}(\sqrt{d})$ and $a + a', aa' \in \mathbb{Z}$, that is, $2x = m \in \mathbb{Z}$, $x^2 - dy^2 = m^2/4 - dy^2 \in \mathbb{Z}$. Due to this last condition, 2 is the only factor of the denominator of y after reducing the fraction. Let us set $y := n/2$. Then $m^2/4 - dy^2 = m^2/4 - d(n^2/4) \in \mathbb{Z}$ if and only if $m^2 - dn^2 \equiv 0 \mod 4$. The case $d \equiv 0 \mod 4$ is impossible. This means that we have to check the cases $d \equiv 1, 2, 3 \mod 4$. When $d \equiv 1 \mod 4$, the congruence assumes the form $m^2 \equiv n^2 \mod 4$. This is equivalent to $m \equiv n \mod 2$, that is, $m = n + 2l$, and we obtain $a = m/2 + (n/2)\sqrt{d} = l + n(1 + \sqrt{d})/2$ with $l, n \in \mathbb{Z}$. If, in the case $d \equiv 2, 3 \mod 4$, the congruence $m^2 - dn^2 \equiv 0 \mod 4$ had a solution with n odd, $d \equiv m^2 \mod 4$ would lead to $d \equiv 0 \mod 4$ for m even and $d \equiv 1 \mod 4$ for odd m. Both these results contradict the choice of d. For n even, one obtains from the congruence $m^2 \equiv 0 \mod 4$ that m is even as well. Then $x = m/2$ and $y = n/2$ are both integers.

One can visualize the elements $a = x + y\sqrt{d}$ of $\mathbb{Q}(\sqrt{d}\,)$ as points in the plane with Cartesian coordinates

$$R(a) = \frac{a + a'}{2} = x$$

and

$$I(a) = \begin{cases} \dfrac{a - a'}{2} = y\sqrt{d} & \text{for} \quad d > 0, \\[2ex] \dfrac{a - a'}{2i} = \dfrac{y\sqrt{d}}{i} & \text{for} \quad d < 0. \end{cases}$$

Geometrically, $a' = x - y\sqrt{d}$, the conjugate of a, corresponds to reflection at the x-axis. The elements of A_d can be visualized as lattice points in the plane.

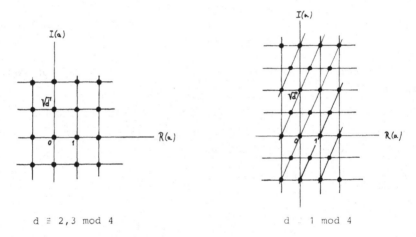

d ≡ 2,3 mod 4   d ≡ 1 mod 4

We have already mentioned that the number theory of the ring A_d corresponds to the theory of binary quadratic forms of discriminant $-d$ or $-d/4$. To see an example of this correspondence we determine the group of units of this ring. Let $x + y\sqrt{d}$ be a unit element. Then $u + v\sqrt{d}$ exists with

$$(x + y\sqrt{d}\,)(u + v\sqrt{d}\,) = 1.$$

It follows that

$$(x - y\sqrt{d}\,)(u - v\sqrt{d}\,) = 1.$$

Consequently, $x - y\sqrt{d}$ is a unit element as well and so is

$$(x + y\sqrt{d}\,)(x - y\sqrt{d}\,) = x^2 - dy^2.$$

Since the only units of $\mathbb{Z}$ are ± 1, we have

$$x^2 - dy^2 = \pm 1.$$

Conversely, if this equation is satisfied, $x + y\sqrt{d}$ is a unit element. Hence there is a bijective correspondence between the units and the solutions of Fermat's (Pell's) equation $x^2 - dy^2 = \pm 1$, where $x, y \in \mathbb{Z}$ for $d \equiv 2, 3$ mod 4 and $2x, 2y, x + y \in \mathbb{Z}$ for $d \equiv 1$ mod 4.

We know the following theorem from previous considerations.

(6.12) **Theorem.** *The units of A_d are*

$1, i, -1, -i$	*for* $d = -1$,
$1, \epsilon, \ldots, \epsilon^5, \quad \epsilon := \exp(2\pi i/6)$	*for* $d = -3$,
$1, -1$	*for* $d < 0, \quad d \neq -1, -3$,
$\pm \epsilon^k, \quad k \in \mathbb{Z}, \quad \epsilon$ *fundamental unit*	*for* $d > 0$.

(A *fundamental unit* ϵ is of the form $\epsilon = x + y\sqrt{d}$, where (x, y) is the smallest nontrivial solution of the corresponding Fermat (Pell) equation).

Units can be visualized as lattice points on the circle: $\{\alpha = x + y\sqrt{d} \mid x, y \in \mathbb{R}, R(\alpha)^2 + I(\alpha)^2 = 1\}$ for $d < 0$ or on the hyperbolas $\{\alpha = x + y\sqrt{d} \mid x, y \in \mathbb{R}, R(\alpha)^2 - I(\alpha)^2 = \pm 1\}$ for $d > 0$.

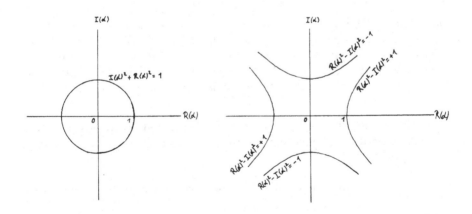

After settling the question of how to define the ring of integers in A_d, one asks oneself: For which d does the theorem of unique prime factorization hold. This question is a natural one also in the language of binary forms because it means: For which d is there only one form of discriminant d or $d/4$ (up to proper equivalence)? We have already seen that this is an important question when representing numbers by quadratic forms. The exact connection will be explained later.

Dirichlet was the first to see clearly that unique prime factorization does not always hold. There are, for example, in A_{-5} these two different decompositions

$$6 = 2 \cdot 3 = (1 + \sqrt{-5})(1 - \sqrt{-5}).$$

Gauss determined, without proof, all negative d for which A_d has the property that A_d is a principal ideal domain. However, it was only recently that the last step in proving this was taken (cf. A. Baker: *Transcendental Number Theory*, Cambridge University Press, Cambridge, England 1975, and H. M. Stark: A complete determination of the complex quadratic fields with class number one, *Michigan Math. J.* **14** (1967), 1–27.) If d is positive we are not even close to solving the problem. There is a conjecture that A_d is a principal ideal domain for infinitely many d; but no promising approach is in sight to prove this.

A natural approach to a solution is to investigate for what d the norm function norm$(a) = N(a) = aa'$ defines a Euclidean algorithm. If it does, unique prime factorization follows. The problem has an easy solution for negative d but is difficult for positive d and was settled only about 1950. Some of these cases are treated in Hasse's *Lectures on Number Theory*.

(6.13) **Theorem.** *Let $d < 0$. Then A_d is Euclidean with respect to its norm only for $d = -1, -2, -3, -7, -11$. So for these values of d we have unique factorization. Moreover, the remaining imaginary quadratic fields with unique factorization correspond to $d = -19, -43, -67, -163$. If $d > 0$, then A_d is Euclidean with respect to its norm for the following values: $d = 2, 3, 5, 6, 7, 11, 13, 17, 19, 21, 29, 33, 37, 41, 57, 73$. So again, we have unique factorization for these d. In addition, unique factorization holds for "many other" d.*

To determine whether A_d is Euclidean one proceeds as follows. One has to determine those d such that for any two numbers $\gamma, \gamma_1 \in A_d$, $\gamma_1 \neq 0$, a $\beta \in A_\alpha$ exists such that

$$|N(\gamma_2)| < |N(\gamma_1)| \quad \text{for} \quad \gamma = \beta\gamma_1 + \gamma_2.$$

This condition is equivalent to the following condition. For any $\alpha \in \mathbb{Q}(\sqrt{d})$ there is $\beta \in A_d$ such that

$$|N(\alpha - \beta)| < 1.$$

We begin with the case $d < 0$ and $d = 2$ or $3 \bmod 4$. Then $A_d = \mathbb{Z} \oplus \mathbb{Z}\sqrt{d}$ and the elements of A_d can be identified with the points of a lattice in the plane. We consider each lattice point as the center of a rectangle R_β parallel to the axes and with edges of length 1 and $\sqrt{d}$.

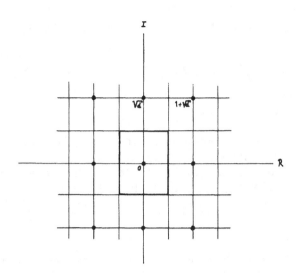

The maximal distance between an arbitrary $\alpha \in \mathbb{Q}(\sqrt{d})$ and a lattice point, say 0, is realized by the corners of the rectangles, that is, e.g., by $\alpha = \frac{1}{2}(1 + \sqrt{d})$ for $\beta = 0$. Thus our necessary and sufficient condition is equivalent to

$$\left| N\left(\tfrac{1}{2}(1 + \sqrt{d}) \right) \right| = \tfrac{1}{4}(1 + |d|) < 1,$$

that is $d = -1$ or -2.

If $d < 0$ and $d = 1 \bmod 4$, then A_d contains not only the lattice points $\mathbb{Z} \oplus \mathbb{Z}\sqrt{d}$ but also the corners of the rectangles R_β (see the figure above). By constructing a suitable hexagon around any element of A_d one sees that points α of maximal distance from any $\beta \in A_d$ are, for example, the corner points on the "imaginary" axis.

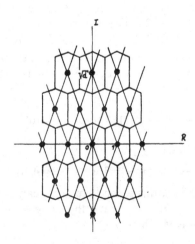

A simple calculation shows that the first such point has the l-coordinate $\frac{1}{4}(\sqrt{d} + 1/\sqrt{d})$. This leads to the condition

$$\left| N\left(\frac{1}{4}\left(1 + \frac{1}{d}\right)\sqrt{d}\right) \right| = \frac{1}{16}\left(|d| + 2 + \frac{1}{|d|}\right) < 1,$$

that is $d = -3, -7,$ or -11.

Let us now give two typical applications of (6.13) to two questions of Fermat and Euler that were mentioned earlier.

First let us consider the equation $y^2 + 2 = x^3$. In (2.7), we claimed that $x = 3$, $y = 5$ is the only solution in natural numbers. We use an idea of Euler to show this. Let x, y be an arbitrary solution. Then, in $\mathbb{Z}[\sqrt{-2}]$,

$$x^3 = y^2 + 2 = (y + \sqrt{-2})(y - \sqrt{-2}).$$

Since there is unique prime factorization in $\mathbb{Z}[\sqrt{-2}]$ and since $(y + \sqrt{-2})$ and $(y - \sqrt{-2})$ are relatively prime (!), $y + \sqrt{-2}$ can by (6.12) be written as a cube (up to sign):

$$y + \sqrt{-2} = \pm\left(x_1 + x_2\sqrt{-2}\right)^3$$
$$= \pm\left(x_1^3 - 2x_1x_2^2 - 4x_1x_2^2\right) + \left(2x_1^2x_2 + x_1^2x_2 - 2x_2^3\right)\sqrt{-2}.$$

Comparing coefficients, one obtains

$$1 = x_2\left(2x_1^2 + x_1^2 - 2x_2^2\right), \qquad y = \pm\left(x_1^3 - 2x_1x_2^2 - 4x_1x_2^2\right),$$

i.e., $x_2 = \pm 1$, $x_1 = +1$, $y = \pm 5$, and hence our statement. The theory for the general equation $y^2 + k = x^3$ with $k \in \mathbb{Z}$ is very interesting; see L. J. Mordell, *Diophantine Equations*, §26.

Let us now continue with another application of (6.13) and sketch a proof of Euler's discovery, mentioned in Chapter 3, that $x^2 + x + 41$ is prime for $x = 0, 1, 2, 3, \ldots, 39$. The expression $x^2 + x + 41$ is the norm of $\frac{1}{2}(2x + 1 + \sqrt{-163})$ in A_{-163} and consequently not a prime element in A_{-163}. Without loss of generality we can assume that the prime factor $u + v\sqrt{-163}$ $(u, v \in \frac{1}{2}\mathbb{Z}, u + v \in \mathbb{Z})$ of $x^2 + x + 41$ is a factor of $\frac{1}{2}(2x + 1 + \sqrt{-163})$ in A_{-163}:

$$\tfrac{1}{2}(2x + 1 + \sqrt{-163}) = (u + v\sqrt{-163})(r + s\sqrt{-163}), \qquad r, s \in \tfrac{1}{2}\mathbb{Z}, r + s \in \mathbb{Z}.$$

Separating real and imaginary parts, one obtains

$$\tfrac{1}{2}(2x + 1) = ru - 163vs, \qquad 2 = 4(us + rv).$$

Save for exceptional cases, this second equation can only hold if us and vr have opposite signs, i.e., ru, $-vs$ have the same signs. In the latter case, $x \geqslant 40\frac{1}{2}$. An exception is when $0 \leqslant x \leqslant 39$. Then $r = \pm 1$, $s = 0$, consequently $\frac{1}{2}(2x + 1 + \sqrt{-163})$ is a prime element in A_{-163}, i.e., $x^2 + x + 41$ is a rational prime for $0 \leqslant x \leqslant 39$.

Let us now interrupt our discussion of A_d to formulate an interesting program of exercises for the reader. Define a Zeta-function for those negative d with unique prime factorization by analogy to our considerations for A_{-1}. This Zeta-function should then be expanded in an Euler product. Next, explicitly determine the prime elements and write the Zeta-function ζ_d for A_d in the form

$$\zeta_d(s) = \mu \zeta(s) L_d(s),$$

where μ is the number of units and $L_d(s)$ a suitable L-series. Then express the residue for $s = 1$, i.e., $\lim_{s \downarrow 1}(s - 1)\zeta_d(s)$, as a series and interpret $\zeta_d(s)$ as a Riemann sum for a suitable double integral. This can be computed and $\lim(s - 1)\zeta_d(s)$ can be calculated. Comparing both results one obtains the following formulas for $-d = 2, 3, 7$.

(6.14) Theorem.

$$1 + \frac{1}{3} - \frac{1}{5} - \frac{1}{7} + \frac{1}{9} + \frac{1}{11} - - + + \cdots = \frac{\pi}{2\sqrt{2}} \qquad (Newton),$$

$$1 - \frac{1}{2} + \frac{1}{4} - \frac{1}{5} + \frac{1}{7} - \frac{1}{8} + - \cdots = \frac{\pi}{3\sqrt{3}} \qquad (Euler),$$

$$1 + \frac{1}{2} - \frac{1}{3} + \frac{1}{4} - \frac{1}{5} - \frac{1}{6} + \frac{1}{8} + \cdots = \frac{\pi}{\sqrt{7}} \qquad (Euler).$$

(We recommend the case $d = -163$ to the indefatigable calculator. If necessary, help may be found in Chapter 8.)

We will now give a systematic development of the connections between the ideal theory of the rings A_d and binary forms. First let us look once again at the diagram below (readers with some knowledge of algebraic number theory know what is significant: $\mathbb{Q}$ is the quotient field of the Dedekind ring $\mathbb{Z}$, and $\mathbb{Q}(\sqrt{d})/\mathbb{Q}$ is a finite separable extension, and A_d is the integral closure of $\mathbb{Z}$ in $\mathbb{Q}(\sqrt{d})$. Specifically, $A_d \cap \mathbb{Q} = \mathbb{Z}$; in our case, A_d is a free $\mathbb{Z}$-module of rank 2.)

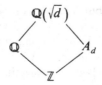

(6.15) **Definition.** A subset $M \subset \mathbb{Q}(\sqrt{d})$ is called a *module* (or *fractional ideal*) if the following conditions are satisfied.

(1) There are finitely many (so-called) generating elements $b_1, \ldots, b_m$ in $\mathbb{Q}(\sqrt{d})$ such that every element of M can be written in the form

$$\alpha_1 b_1 + \cdots + \alpha_m b_m \qquad \text{with} \quad \alpha_i \in \mathbb{Z}.$$

(2) If $a \in A_d$ and $x \in M$, then $ax \in M$.

EXAMPLES.

(1) A_d is a module.
(2) Let $M \subset A_d$. Then M is a module if and only if M is an ideal of A_d. (One can show that every ideal is finitely generated.)
(3) $\mathbb{C}(\sqrt{d})$ is not a module because $\mathbb{Q}(\sqrt{d})$ does not satisfy condition (1) of our definition.

(6.16) **Remark.** Every nontrivial module M not only is finitely generated but has a basis with two elements.

PROOF. This proof will use the theory of finitely generated abelian groups. Every (nontrivial) module M is a finitely generated abelian group. Since $M \subset \mathbb{Q}(\sqrt{d})$, M is torsion free and is therefore a free abelian group, i.e., there is a basis. If the number of elements $c_1, \ldots, c_m$ in a basis were $\geqslant 3$, $c_1, \ldots, c_m$ would then be linearly dependent over $\mathbb{Q}$ since they belong to a two-dimensional $\mathbb{Q}$-vector space. Then $0 = \alpha_1 c_1 + \cdots + \alpha_m c_m$ with $\alpha_i \in \mathbb{Q}$ and not all the α_i vanishing; multiplying this equation by the denominators of α_i shows that $c_1, \ldots, c_m$ would be linearly dependent over $\mathbb{Z}$ which contradicts the fact that the c_i form a basis. $m = 1$ is equally impossible; for in that case, every element of M could uniquely be written in the form αc with $\alpha \in \mathbb{Z}$ and $c \in M - \{0\}$; specifically, for the element ωc in M, we have $\omega c = \alpha c$, whence $\sqrt{d} \in \mathbb{Z}$. This is a contradiction.

Modules can be visualized in a natural way as lattices in an appropriate plane.

Taking the norm map establishes a correspondence between quadratic number fields and binary quadratic forms. Let $M = \langle a, b \rangle := a\mathbb{Z} \oplus b\mathbb{Z}$ be a module and $ax + by$, $x, y \in \mathbb{Z}$ be an element in M. Then

$$(ax + by)(a'x + b'y) = aa'x^2 + (ab' + a'b)xy + bb'y^2$$

is the norm of this element. Usually, the values of this binary quadratic form in x and y are rational but not integral. However, we are interested in integral quadratic forms. We want to associate to every integral binary quadratic form (in a way that has to be made more precise) a module in a quadratic number field and vice versa. We have just seen that this correspondence can not be established by restricting the field norm to the given module and using the resulting quadratic form. A more careful analysis is needed.

Let us compare the modules $A_d = 1\mathbb{Z} \oplus \omega\mathbb{Z}$ and $M = a\mathbb{Z} \oplus b\mathbb{Z}$. Let $\sigma : \mathbb{Q}(\sqrt{d}) \to \mathbb{Q}(\sqrt{d})$ be a $\mathbb{Q}$-linear mapping with $\sigma(1) = a$ and $\sigma(\omega) = b$, i.e., $\sigma(A_d) = M$. The condition $\sigma(A_d) = M$ defines the determinant $\det(\sigma)$ up to sign; for letting τ be another mapping with $\tau(A_d) = M$, then $\tau^{-1} \circ \sigma$ is an automorphism of A_d. This means that it can be described by an element of $\mathrm{GL}(2, \mathbb{Z})$ and consequently has determinant ± 1. We can therefore define Norm $(M) := N(M) := |\det(\sigma)|$.

For example, let us consider the element $h = \alpha + \beta\sqrt{d} \in \mathbb{Q}(\sqrt{d})$, $h \neq 0$, and the module $hA_d = h\mathbb{Z} \oplus h\omega\mathbb{Z}$. In this case, σ can be described by the matrices

$$\begin{bmatrix} \alpha & \beta \\ \beta d & \alpha \end{bmatrix} \quad \text{or} \quad \begin{bmatrix} \alpha - \beta & 2\beta \\ \dfrac{\beta d - \beta}{2} & \alpha + \beta \end{bmatrix}$$

for $d \equiv 2, 3 \bmod 4$ or $d \equiv 1 \bmod 4$. Then

$$N(hA_d) = |\det(\sigma)| = |\alpha^2 - \beta^2 d| = |N(h)|,$$

i.e., the norm of the module hA_d is given by the norm of the element h. More generally:

(6.17) Remark. Let a module M and $h \in \mathbb{Q}(\sqrt{d})$, $h \neq 0$ be given. Then

$$N(hM) = |N(h)|N(M),$$

for it is easy to see that $\det(A_d \to hM) = \det(A_d \to hA_d)\det(A_d \to M)$, i.e., $\det(hA_d \to hM) = \det(A_d \to M)$.

Can one say something in general about the norm of a module? To answer this question we may assume that our module M is of the form $M = \langle 1, c \rangle = \mathbb{Z} \oplus \mathbb{Z}c$ because later we will not need to distinguish between modules M_1, M_2 that are constant multiples of each other, i.e., without loss of generality, we can replace $\langle a, b \rangle$ by $(1/a)\langle a, b \rangle = \langle 1, b/a \rangle =: \langle 1, c \rangle$. Now we will show that c is a root of a polynomial

$$rt^2 + kt + l$$

with $r, k, l \in \mathbb{Z}$, $r > 0$, g.c.d$(r, k, l) = 1$. To see this, we have only to consider the minimal polynomial $(t - c)(t - c') = t^2 - (c + c')t + cc'$, form the greatest common denominator r (or its negative) of $c + c'$ and cc', and write

$$t^2 - (c + c')t + cc' = t^2 + \frac{k}{r}t + \frac{l}{r}$$

with g.c.d$(r, k, l) = 1$. Then c is a root of the polynomial

$$\phi_c(t) := rt^2 + kt + l$$

which is uniquely determined by these conditions. (Obviously $\phi_c = \phi_{c'}$ and $\phi_{c_0} = \phi_c$ if and only if $c_0 = c$ or $= c'$.) This leads to

(6.18) Remark. $N(M) = 1/r$.

PROOF. To prove this it suffices to show that $A_d = \langle 1, rc \rangle$ for then $\begin{pmatrix} 1 & 0 \\ 0 & 1/r \end{pmatrix}$ is the matrix which effects the transition from the basis $1, rc$ to the basis $1, c$. This matrix has determinant $1/r$ which is the norm of M. First we show $A_d \subset \langle 1, rc \rangle$. Let $a \in A_d$. Then $a \langle 1, c \rangle \subset \langle 1, c \rangle$, i.e., $a \cdot 1 = x + yc$ and $a \cdot c = x_1 + y_1 c$ with $x, y, x_1, y_1 \in \mathbb{Z}$. It follows that $xc + yc^2 = x_1 + y_1 c$. Since $c^2 = -(k/r)c - l/r$, we have $xc + y(-(k/r)c - l/r) = -yl/r + (x - ky/r)c = x_1 + y_1 c$. Hence $x_1 = -yl/r \in \mathbb{Z}$ and $y_1 = x - ky/r \in \mathbb{Z}$. Since g.c.d$(r, k, l) = 1$, it follows that r divides y. Let $y = y_0 r$ with $y_0 \in \mathbb{Z}$. Then $a = x + yc = x + y_0 rc \in \langle 1, rc \rangle$.

Conversely, we show that $\langle 1, rc \rangle \subset A_d$. It suffices to show that $r, c \in A_d$. rc is a root of the polynomial $t^2 + kt + lr$ with $k = rc + rc'$, $lr = (rc)(rc')$. $rc \in A_d$ follows from the characterization of A_d in (6.11).

In the same way as above one shows that $\langle 1, c \rangle = \mathbb{Z}1 \oplus \mathbb{Z}c$ determines a module in a quadratic number field for every root of a polynomial $rt^2 + kt + l$ with $r, k, l \in \mathbb{Z}$ and g.c.d$(r, k, l) = 1$.

(6.19) Definition. Two modules M_1 and M_2 in a quadratic number field $\mathbb{Q}(\sqrt{d})$ are called *equivalent* if there is an element $h \in \mathbb{Q}(\sqrt{d})$, $h \neq 0$, such that $M_1 = hM_2$.

This does indeed define an equivalence relation. We have already seen that every module is equivalent to a module of the form $\langle 1, c \rangle$.

We consider the correspondence between a basis (a, b) of a module M in $\mathbb{Q}(\sqrt{d})$ and the binary quadratic form

$$q_{a,b}(x, y) := \frac{N(ax + by)}{N(M)}.$$

We will show that this form has integral coefficients; consequently, it is a good candidate for the desired correspondence. If one changes the basis, the corresponding forms are equivalent. If one replaces $M = \langle a, b \rangle$ by an equivalent module hM then, by applying (6.17),

$$q_{ha,hb}(x, y) = \frac{N(hax + hby)}{N(hM)}$$

$$= \frac{N(h)}{|N(h)|} \frac{N(ax + by)}{N(M)}$$

$$= \pm q_{a,b}(x, y).$$

This leads to the following definition.

(6.20) Definition. Two modules M_1, M_2 in $\mathbb{Q}(\sqrt{d})$ are called *properly equivalent* if there is an element $h \in \mathbb{Q}(\sqrt{d})$ with $N(h) > 0$ such that $M_1 = hM_2$.

If we replace the module $\langle a, b \rangle$ by a properly equivalent module $\langle ha, hb \rangle$, then $q_{a,b} = q_{ha,hb}$. For $d < 0$, the norm $N(h) = \alpha^2 - d\beta^2$ is > 0 for every $h = \alpha + \beta\sqrt{d} \in \mathbb{Q}(\sqrt{d}) - \{0\}$. This means that the concepts equivalent and properly equivalent are the same for an imaginary quadratic field $\mathbb{Q}(\sqrt{d})$. In the real quadratic case $d > 0$, they coincide if there is a unit ϵ with norm -1. Since $\epsilon M_2 = M_2$ for an arbitrary module M_2 in $\mathbb{Q}(\sqrt{d})$, two equivalent modules M_1, M_2, say $M_1 = hM_2$ with $N(h) < 0$, can be made properly equivalent by replacing h by ϵh. Incidentally, the converse is true also, i.e., if equivalence of two modules in $\mathbb{Q}(\sqrt{d})$ is the same as proper equivalence then there is a unit with norm -1. In the real quadratic case, proper equivalence differs from equivalence only if all units of A_d have norm 1. It is clear that in this case every equivalence class of modules contains exactly two proper equivalence classes.

(6.21) **Remark.** $q_{a,b}$ is a primitive integral quadratic form of discriminant $-d$ if $d \equiv 2, 3 \mod 4$ and of the discriminant $-d/4$ if $d \equiv 1 \mod 4$. If $d < 0$ then $q_{a,b}$ is positive.

A quadratic form $ax^2 + bxy + cy^2$ is *primitive* if the greatest common divisor of the coefficients a, b, c is equal to 1. Obviously, this concept is compatible with equivalence of forms. The discriminant of $ax^2 + bxy + cy^2$ is defined by the determinant of $\left(\begin{smallmatrix} a & b/2 \\ b/2 & c \end{smallmatrix}\right)$, that is, by $ac - b^2/4$.

For a proof of this remark we assume without loss of generality that the module M has the form $M = \langle 1, c \rangle$. This is legitimate because $\langle a, b \rangle$ is equivalent to $\langle 1, c \rangle$ with $c := b/a$; also, $q_{a,b} = \pm q_{1,c} = q_{1,c}$ for $d < 0$. Using the notation of page 85 and (6.18), one obtains

$$q_{1,c}(x, y) = \frac{N(x + cy)}{N(M)} = \frac{x^2 + (c + c')xy + cc'y^2}{N(M)}$$

$$= r\left(x^2 - \frac{k}{r}xy + \frac{l}{r}y^2\right)$$

$$= rx^2 - kxy + ly^2.$$

All values of $q_{1,c}$ are obviously in $\mathbb{Z}$ and g.c.d.$(r, k, l) = 1$, i.e., $q_{1,c}$ is primitive. If $d < 0$ then $q_{1,c}$ is positive because $N(\alpha) \geqslant 0$ for all $\alpha \in \mathbb{Q}(\sqrt{d})$. In addition, $\det(q_{1,c}) = \det(rN(x + yc)) = r^2\det(N(x + yc)) = (r^2/r^2)\det(N(x + yrc)) = \det(q_{1,\omega}) = -d$, if $d \equiv 2, 3 \mod 4$, $= -d/4$, if $d \equiv 1 \mod 4$.

(6.22) **Theorem.** *The set of equivalence classes (proper equivalence classes) of modules in $\mathbb{Q}(\sqrt{d})$ forms an abelian group with respect to the product*

$$M_1 M_2 := \langle \{\alpha\beta \mid \alpha \in M_1, \beta \in M_2\}\rangle.$$

PROOF. Let $M_1 = \langle a, b \rangle$, $M_2 = \langle c, d \rangle$. Then $M_1 M_2$ is the set of all $\mathbb{Z}$-linear combinations of ac, ad, bc, bd; hence it is finitely generated. With $\alpha \in A_d$,

$\alpha M_1 M_2 \subset M_1 M_2$. This means that $M_1 M_2$ is a module. Obviously, the associative and commutative laws are satisfied. The module A_d is the neutral element. The inverse to $\langle a, b \rangle$ is given by $\langle a', b' \rangle$, where a', b' are the conjugates of a, b. To show this, we notice that $\langle a, b \rangle \langle a', b' \rangle$ is equivalent to $\langle 1, c \rangle \langle 1, c' \rangle$ with $c = b/a$, and this module is generated by $1, c, c', cc'$, and consequently by $1, c + c', cc', c$ as well. c is a root of $\phi_c(t) = rt^2 + kt + l$, where $r > 0$ and g.c.d$(r, k, l) = 1$ (see page 86). Consequently, $\langle 1, c \rangle \langle 1, c' \rangle$ will also be generated by $1, k/r, l/r, c$. Also, $\langle 1, c \rangle \langle 1, c' \rangle$ is equivalent to $N(\langle 1, c \rangle)^{-1} \langle 1, c \rangle \langle 1, c' \rangle$; because $N(\langle 1, c \rangle) = l/r$ (see (6.18)) it is equivalent to the module generated by r, k, l, rc or $1, rc$. By the proof of (6.18), $\langle 1, rc \rangle = A_d$.

From now on, we will only consider bases a, b with

$$\det \begin{pmatrix} a & b \\ a' & b' \end{pmatrix} > 0 \qquad \text{for} \quad d > 0,$$

$$i \det \begin{pmatrix} a & b \\ a' & b' \end{pmatrix} > 0 \qquad \text{for} \quad d < 0 \tag{$*$}$$

for modules M in $\mathbb{Q}(\sqrt{d})$. One can always find such bases, if necessary, by changing the order of the basis elements. Geometrically, the condition $(*)$ is an orientation. This orientation becomes relevant when we define a mapping from the set of proper equivalence classes of modules into the set of proper equivalence classes of primitive binary quadratic forms (for fixed d). For if one assigns to every basis $\langle a, b \rangle$ (oriented as above) of a module in $\mathbb{Q}(\sqrt{d})$ the form

$$q_{a,b}(x, y) = \frac{N(ax + by)}{N(M)},$$

then this is a well-defined mapping ψ of the set of proper equivalence classes of modules in $\mathbb{Q}(\sqrt{d})$ into the set of proper equivalence classes of primitive binary quadratic forms with "correct" discriminant. If M_1, M_2 are properly equivalent, say $M_1 = hM_2$ with $N(h) > 0$ and if a_1, a_2 and b_1, b_2 are oriented bases in M_1 and M_2, then there is a matrix $\begin{pmatrix} \alpha & \beta \\ \gamma & \delta \end{pmatrix} \in \mathrm{GL}(2, \mathbb{Z})$ such that

$$hb_1 = \alpha a_1 + \beta a_2, hb_2 = \gamma a_1 + \delta a_2. \tag{1}$$

Then

$$(\alpha\delta - \beta\gamma)(a_1 a_2' - a_1' a_2) = hh'(b_1 b_2' - b_1' b_2)$$
$$= N(h)(b_1 b_2' - b_1' b_2). \tag{2}$$

Since $N(h) > 0$ and $(*)$, we have $\alpha\delta - \beta\gamma = +1$. Together with (1) and (6.17) one easily finds that q_{a_1, a_2} and q_{b_1, b_2} are properly equivalent.

(6.23) **Theorem.** *The mapping ψ is a bijection between the set of proper equivalence classes of modules in $\mathbb{Q}(\sqrt{d})$ and the set of proper equivalence* *classes of primitive positive forms of discriminant $-d$ or $-d/4$ for $d < 0$. If $d > 0$ then ψ is a bijection between the set of proper equivalence classes of modules in $\mathbb{Q}(\sqrt{d}\,)$ and the set of proper equivalence classes of primitive forms of discriminant $-d$ or $-d/4$. The discriminant $-d$ occurs when $d \equiv 2, 3$ mod 4 and $-d/4$ occurs when $d \equiv 1$ mod 4.*

The group of equivalence classes (or of proper equivalence classes) of modules in $\mathbb{Q}(\sqrt{d}\,)$ is called the *class group (narrow class group)* of $\mathbb{Q}(\sqrt{d}\,)$. Then:

(6.24) Corollary. *The (narrow) class group of $\mathbb{Q}(\sqrt{d}\,)$ is finite.*

This follows easily from the fact that the set of proper equivalence classes of binary quadratic forms of fixed discriminant is finite (see Chapter 4). The order h (or $\bar{h}$) of the class group (narrow class group) of $\mathbb{Q}(\sqrt{d}\,)$ is called the *class number (narrow class number)* of $\mathbb{Q}(\sqrt{d}\,)$.

Let us summarize:

$$\bar{h} = h \qquad \text{for} \quad d < 0,$$

$$\bar{h} = h \qquad \text{for} \quad d > 0, \quad \text{if a unit } \epsilon \text{ exists with } N(\epsilon) = -1,$$

$$\bar{h} = 2h \qquad \text{for} \quad d > 0, \quad \text{if } N(\epsilon) = +1 \text{ for all units } \epsilon.$$

From (6.23) we immediately obtain:

(6.25) Corollary. A_d *is a principal ideal domain if and only if the class number h of $\mathbb{Q}(\sqrt{d}\,)$ is 1. A_d is a principal ideal domain with the additional property that each ideal contains a generator with positive norm if and only if there exists one class of properly equivalent primitive (positive for $d < 0$) quadratic forms with discriminant $-d$ or $-d/4$.*

PROOF OF (6.23). The mapping ψ is surjective: Let $rx^2 + kxy + ly^2$ be a primitive form with "correct" discriminant. Let $M := \langle 1, c \rangle$, where c is a root of $rx^2 + kx + l$ such that $\beta < 0$ in the representation $c = \alpha + \beta\sqrt{d}$ (if c does not have this property, c' will). Then $1, c$ is an oriented basis in the sense of (*) and we have $q_{1,c}(x, y) = rN(x + cy) = rx^2 + kxy + ly^2$.

The mapping ψ is injective: Let $M_1 = \langle 1, a \rangle$, $M_2 = \langle 1, b \rangle$ be two modules oriented in the sense of (*) and $q_{1,a}, q_{1,b}$ be properly equivalent by $\left(\begin{smallmatrix} \alpha & \beta \\ \gamma & \delta \end{smallmatrix}\right)$ $\in \mathrm{SL}(2, \mathbb{Z})$. Then

$$q_{1,a}(\alpha x + \beta y, \gamma x + \delta y) = q_{1,b}(x, y)$$

which can be written as

$$\frac{1}{N(M_1)}\left((\alpha + \gamma a)x + (\beta + \delta a)y\right) \cdot \left((\alpha + \gamma a')x + (\beta + \delta a')y\right)$$

$$= \frac{1}{N(M_2)}(x + by)(x + b'y). \tag{3}$$

The numbers $-b$ and $-b'$ are the roots of $q_{1,b}(x, 1) = q_{1,a}(\alpha x + \beta,$ $\gamma x + \delta)$. The roots of the last term are $-(\beta + \delta a)/(\alpha + \gamma a)$ and its conjugate. Consequently,

$$\frac{\alpha + \gamma a}{\beta + \delta a} = \frac{1}{b} \quad \text{or} \quad \frac{1}{b'} .$$

This means that $h \in \mathbb{Q}(\sqrt{d})$ exists such that

$$\alpha + \gamma a = h,$$
$$\beta + \delta a = hb \quad \text{or} \quad = hb'.$$

From (3) it follows that $0 < N(M_1)N(M_2)^{-1} = hh' = N(h)$. If we assume that $\beta + \delta a = hb'$, then, similar to (2),

$$(\alpha\delta - \beta\gamma)(a' - a) = -hh'(b' - b).$$

This is impossible because the bases are oriented in the sense of (*). Consequently, h, hb is a basis of M_1, i.e., $M_1 = \langle h, hb \rangle = h\langle 1, b \rangle = hM_2$. M_1 is properly equivalent to M_2. Using remark (6.21), this completes our proof.

Gauss formulated these propositions in the language of binary quadratic forms which is often quite complicated. The so-called "composition" of forms corresponds to the multiplication of modules. The group which arises in this way seems to be one of the first examples of a non-obvious group structure, that is, one not arising from a permutation group.

When $d < 0$, one determines the number of primitive reduced forms of discriminant $-d$ or $-d/4$ in order to find the class number h of $\mathbb{Q}(\sqrt{d})$. This means that one has to count the number of triples (a, b, c) with $|b| \leqslant a \leqslant c$; $-a < b \leqslant a$; $0 \leqslant b \leqslant a$, if $a = c$; $ac - b^2/4 = -d$ or $= -d/4$ and g.c.d$(a, b, c) = 1$; cf. (4.2).

EXAMPLES. (1) $d = -23$. Since $-23 \equiv 1 \mod 4$ we have $ac - b^2/4 = 23/4$. Our conditions allow three possibilities:

$$\begin{pmatrix} 1 & \frac{1}{2} \\ \frac{1}{2} & 6 \end{pmatrix}, \quad \begin{pmatrix} 2 & \pm\frac{1}{2} \\ \pm\frac{1}{2} & 3 \end{pmatrix},$$

and consequently $h = 3$.

(2) $d = -47$. Since $-47 \equiv 1 \mod 4$ we have $ac - b^2/4 = 47/4$. Our conditions allow the possibilities

$$\begin{pmatrix} 1 & \frac{1}{2} \\ \frac{1}{2} & 12 \end{pmatrix}, \quad \begin{pmatrix} 2 & \pm\frac{1}{2} \\ \pm\frac{1}{2} & 6 \end{pmatrix}, \quad \begin{pmatrix} 3 & \pm\frac{1}{2} \\ \pm\frac{1}{2} & 4 \end{pmatrix}.$$

Consequently, $h = 5$.

We now apply (6.25) to prove that A_d is a principal ideal domain for certain $d > 0$. When $d = 2$, all reduced binary quadratic forms are given by

$\left(\begin{smallmatrix}-1 & 0\\ 0 & 2\end{smallmatrix}\right)$ and $\left(\begin{smallmatrix}1 & 0\\ 0 & -2\end{smallmatrix}\right)$ (see page 41). Using $\left(\begin{smallmatrix}1 & 1\\ -2 & -1\end{smallmatrix}\right) \in SL(2, \mathbb{Z})$, these are properly equivalent. Therefore, every module in $\mathbb{Q}(\sqrt{2})$ is properly equivalent to A_2; specifically, A_2 is a principal ideal domain.

When $d = 3$, $\left(\begin{smallmatrix}-1 & 0\\ 0 & 3\end{smallmatrix}\right), \left(\begin{smallmatrix}1 & 0\\ 0 & -3\end{smallmatrix}\right)$ are the only reduced forms. They are not equivalent because -1 can be represented by $-x^2 + 3y^2$ but not by $x^2 - 3y^2$. By (6.23), every module will be properly equivalent to the modules belonging to these forms, i.e., to $\langle 1,\sqrt{3}\rangle = A_3$ or $\langle 1,\sqrt{1/3}\rangle$. Clearly, $3\langle 1,\sqrt{1/3}\rangle = \langle 3,\sqrt{3}\rangle$; this last module is equivalent to $\langle 1,\sqrt{3}\rangle = A_3$ using the factor $\sqrt{3}$. Hence A_3 is a principal ideal domain.

When $d = 5$, the only primitive reduced forms with determinant $-5/4$ are

$$\begin{pmatrix} 1 & \frac{1}{2} \\ \frac{1}{2} & -1 \end{pmatrix}, \quad \begin{pmatrix} -1 & \frac{1}{2} \\ \frac{1}{2} & 1 \end{pmatrix}.$$

These forms are equivalent using $\left(\begin{smallmatrix}0 & 1\\ 1 & 0\end{smallmatrix}\right)$. Since the equation $x^2 - 5y^2 = -1$ has a nontrivial solution, these forms are properly equivalent (see page 87). Hence A_5 is a principal ideal domain. In a similar way, one shows for many other positive d that A_d is principal, e.g., for $d = 6, 7, 11, 13, 14, 17, 19, 21, 22, 23, 29, 31$.

The class number will also be studied in the following chapter when we sketch an analytical proof for a general formula.

The following theorem, which can be proved purely algebraically, is the first result about the structure of the narrow class group of a quadratic number field:

(6.26) **Theorem.** *Let r be the number of the prime divisors of the discriminant of $\mathbb{Q}(\sqrt{d})$. Then the narrow class group of $\mathbb{Q}(\sqrt{d})$ has exactly 2^{r-1} elements of order $\leqslant 2$.*

Corollary. *In the decomposition of the narrow class group as a product of cyclic groups, exactly $r - 1$ factors of even order occur.*

This theorem is closely tied to the so-called genus theory, one of the most difficult parts of *Disquisitiones Arithmeticae*. Hasse's book, *Number Theory* (III, 26, 8) contains a proof which uses quadratic reciprocity and Dirichlet's theorem about primes in arithmetic progressions.

Let us briefly consider the case $d < 0$ and show that there are at least 2^{r-1} elements of order 2. Without loss of generality, we can assume that d is squarefree. If for example, $d \equiv 2 \bmod 4$ and if $d = ab$ is a decomposition into two relatively prime factors, then an element of order 2 corresponds to the form $ax^2 - by^2$ in the narrow class group. This is so because one has

$$M^2 = \mathbb{Z} + \mathbb{Z}\frac{b}{a} + \mathbb{Z}\sqrt{\frac{b}{a}} = \frac{1}{a}(\mathbb{Z}a + \mathbb{Z}b + \mathbb{Z}\sqrt{d}) = \frac{1}{a}(\mathbb{Z} \oplus \mathbb{Z}\sqrt{d}) = \frac{1}{a}A_d$$

because g.c.d.$(a, b) = 1$ for the module $M = \langle 1,\sqrt{b/a}\rangle$ which belongs to this form. The decomposition of d into r different prime numbers, $d = -p_1 \cdots p_r$, yields at least 2^{r-1} different reduced primitive forms $ax^2 + cy^2$, $ac = -d$, $a < c$. Consequently, the class group contains at least 2^{r-1} elements of order $\leqslant 2$.

The following theorem is obtained by using results from the reduction theory of positive binary quadratic forms (see (4.2)).

(6.27) **Theorem.** *The form which corresponds to the module $\langle 1, c \rangle$ in an imaginary quadratic number field $\mathbb{Q}(\sqrt{d})$ is reduced if and only if the following conditions are satisfied for c: $c > 0$, $-\frac{1}{2} < \operatorname{Re} c \leqslant \frac{1}{2}$, $|c| > 1$ for $-\frac{1}{2} < \operatorname{Re} c < 0$, $|c| \geqslant 1$ for $0 \leqslant \operatorname{Re} c \leqslant \frac{1}{2}$.*

Geometrically, these conditions mean that c lies in the domain G in the figure below (the heavy part of the boundary, including i, is part of G; the rest of the boundary is not). This figure which can often be seen in books on function theory can be found in Gauss's posthumously published papers. However, Gauss developed it in a different context. In the real quadratic case $d > 0$, every module M in $\mathbb{Q}(\sqrt{d})$ is equivalent to a module of the form $\langle 1, \theta \rangle$, where $\theta \in \mathbb{Q}(\sqrt{d}) - \mathbb{Q}$ has a periodic expansion into a continued fraction (see (4.17)). We remind the reader (see (4.19)) that θ has a purely periodic expansion if and only if θ is reduced, i.e., if $1 < \theta$, $-1 < \theta' < 0$.

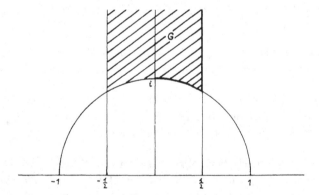

(6.28) **Remark.** A module $\langle 1, \theta \rangle$ in $\mathbb{Q}(\sqrt{d})$, $d > 0$ is equivalent to a module $\langle 1, \theta^* \rangle$, where θ^* is reduced.

One can prove this remark within the framework of the reduction theory of quadratic forms. We prefer a different technique and remind the reader that $\langle 1, a \rangle$ is equivalent to $\langle 1, b \rangle$ if and only if there is a matrix $\left(\begin{smallmatrix} \alpha & \beta \\ \gamma & \delta \end{smallmatrix} \right) \in GL(2, \mathbb{Z})$ such that

$$a = \frac{\alpha b + \beta}{\gamma b + \delta}.$$

This statement already occurs in the proof of (6.23) and can easily be verified.

PROOF OF THE REMARK. Let us use the notation of Chapter 4, page 44. Then

$$\theta = \theta_0 = a_0 + \frac{1}{\theta_1}, \qquad \theta_1 = \frac{1}{\theta - a_0}.$$

Applying $\begin{pmatrix} 0 & 1 \\ 1 & -a_0 \end{pmatrix}$, one sees that $\langle 1, \theta_1 \rangle$ is equivalent to $\langle 1, \theta \rangle$. Let θ_i be an arbitrary number in the expansion of θ. Then $\langle 1, \theta_i \rangle$ is equivalent to $\langle 1, \theta \rangle$. After finitely many steps we find a θ_i with a purely periodic expansion into a continued fraction. Consequently, it is reduced.

The following corollary reduces the question of the equivalence of two modules in a real quadratic field to a strictly computational exercise.

(6.29) Theorem. *Let* $\theta, \theta^* \in \mathbb{Q}(\sqrt{d}) - \mathbb{Q}$, $d > 0$; *let* θ, θ^* *be reduced. The modules* $\langle 1, \theta \rangle$ *and* $\langle 1, \theta^* \rangle$ *are equivalent if and only if* θ^* *occurs in the expansion of* θ *(as one of the* θ_i).

PROOF. Let us assume, θ^* occurs as some θ_i in the expansion into a continued fraction of θ:

$$\theta = \frac{\theta_i p_{i-1} + p_{i-2}}{\theta_i q_{i-1} + q_{i-2}}.$$

Then the modules $\langle 1, \theta \rangle$ and $\langle 1, \theta^* \rangle$ are equivalent because the matrix

$$\begin{pmatrix} p_{i-1} & p_{i-2} \\ q_{i-1} & q_{i-2} \end{pmatrix}$$

is in $\mathrm{GL}(2, \mathbb{Z})$ as $p_{i-1}q_{i-2} - p_{i-2}q_{i-1} = (-1)^i$ (see (4.16')).

To prove the converse we recall that the reduction theory (see page 37) says that $\mathrm{GL}(2, \mathbb{Z})$ can be generated by $\begin{pmatrix} 1 & a \\ 0 & 1 \end{pmatrix}, \begin{pmatrix} 0 & 1 \\ 1 & 0 \end{pmatrix}$, $a \in \mathbb{Z}$ (even by $\begin{pmatrix} 1 & 1 \\ 0 & 1 \end{pmatrix}, \begin{pmatrix} 0 & 1 \\ 1 & 0 \end{pmatrix}$). Since we have assumed that

$$\theta^* = \frac{\alpha\theta + \beta}{\gamma\theta + \delta} \qquad \text{with} \qquad \begin{pmatrix} \alpha & \beta \\ \gamma & \delta \end{pmatrix} \in \mathrm{GL}(2, \mathbb{Z}),$$

it suffices to show that

$$\theta, \theta + 1 = \frac{1\theta + a}{0\theta + 1}, \qquad \frac{1}{\theta} = \frac{0\theta + 1}{1\theta + 0}$$

have the same periods, up to a cyclic permutation. $\theta + a$ satisfies this condition trivially. It consequently suffices to consider $1/(\theta - a_0) = \theta_1$ instead of $1/\theta$; the statement obviously follows.

Let us conclude this chapter with a few words about Gauss's life and personality. In his "Lobrede auf Herrn Leonhard Euler" (Laudatio of Leonhard Euler), Nicolaus Fuss, a distant relative and colleague of Euler, has given a description of what one would expect from a scientific biography. His oration was delivered to a session of the Imperial Academy of Sciences at St. Petersburg on October 23, 1783:

> Whoever describes the life of a great man who has enlightened his century praises the human mind. Nobody should undertake the presentation of such an interesting picture who does not combine a most pleasant style, essential to the orator, with the most perfect knowledge of the sciences whose progress he is to report. Many claim that these two things are not compatible with each other. Even though the biographer need not adorn his subject by unnecessary decorations, this does not release him from the obligation to organize his facts tastefully, to present them clearly, and to narrate them with dignity. He ought to show how Nature brings forth great men; he ought to investigate the circumstances that help in the development of excellent talents; and in his extensive explanations of the learned works of the man whom he praises he must not forget to describe its state before this man appeared and to determine the level from which he started.

Of course, in these few pages we cannot satisfy this program. There is a relatively comprehensive literature about Gauss which gives more information. A new excellent biography, by W. K. Bühler, has just been published. For a shorter summary we refer to K. O. May's article in the *Dictionary of Scientific Biography* and to Maier–Leibniz, "Kreativität," in the Abhandlungen der Braunschweigischen Wissenschaftlichen Gesellschaft, Gauss-Festschrift, 1977.

Gauss was born in Braunschweig on April 30, 1777. His father, Gebhard Dietrich Gauss, worked at several jobs, as a mason, butcher, gardener, and water worker. Before her marriage, his mother was a domestic. His father was always busy trying to improve the poor circumstances of his family, but he was a rather hard and strict man. In a letter of April 15, 1810 to his bride, Minna Waldeck, Gauss writes about his father:

> My father was a perfectly honest and in a way respectable and well-respected man, but at home he was very authoritarian, rough and unrefined; I might well say that he never had my full confidence as a child even though this never led to actual misunderstandings because early on I became very independent of him.

Initially Gauss received little encouragement in his modest domestic arrangements although his uncommon genius manifested itself at an early age. He taught himself how to calculate and read. In elementary school, his teachers, particularly the assistant teacher, Martin Bartels, became aware of Gauss's talent during their arithmetic lessons. He was nine when Bartels started given him special instruction, providing special textbooks and drawing the attention of influential persons to his exceptional student. In 1788, against his father's wishes, Gauss entered the gymnasium, a type of high school. He made such good progress that he was promoted to the top class within two years. In addition to his mathematical talents, his exceptional gift for languages was recognized. In 1791, at the age of 14, Gauss

Carl Friedrich Gauss

was presented at Court. Duke Carl Wilhelm Ferdinand of Braunschweig bestowed a scholarship on him which relieved him from the financial confinements of his home and made it much easier for him to continue his education. This scholarship was regularly renewed until Gauss was 30. Repeatedly, Gauss expressed his deep gratitude which he felt towards his prince. In 1792, Gauss entered the Collegium Carolinum in Braunschweig. At that time, he read the great mathematical classics, among them Newton's *Principia*, J. Bernoulli's *Ars Conjectandi*, and the works of Euler and Lagrange. This was also the time when Gauss started his own research. Three years later, Gauss left the Collegium and began studying in Götting-en, without yet having decided whether to specialize in mathematics or the classical languages. The discovery of the constructability of the regular 17-gon with ruler and compass prompted him to choose mathematics as his vocation. He now entered a phase during which he had so many ideas that he hardly had time to write everything down. His scientific diary is a testimonial to the extent of his research. This period, until 1800, was one of the most fruitful of his life. He never fully developed many results that could have been among his most important works, especially his investigations of elliptic functions. This is also the period during which *Disquisitiones Arithmeticae* was written, a book that was published in 1801 and which, though not much of it was understood when published, catapulted him to prominence as a mathematician. In 1798, Gauss completed his studies and returned to Braunschweig. He continued his research, still benefitting from his scholarship. In 1799, he was, in absentia, given the doctoral degree by the University of Helmstedt. In 1801, Gauss made the calculations which led to the rediscovery of the planetoid Ceres. This made his name well known to the public. Piazzi had found this planetoid in the winter of 1800/1801 but soon lost it. Gauss achieved the seemingly impossible. On the basis of only a few observations and as a consequence of extensive theoretical and numerical investigations, he succeeded in computing its orbit. (In 1978, Ceres made the papers again when a radio signal, reflected by Ceres, was recaptured.) This success led Gauss to turn to astronomy. He involved himself systematically in this science and was made director of the Göttingen Observatory in 1807, a position which he retained until his death in 1855. This allowed him to pursue his research largely independently of any teaching obligations, but it also entailed much practical work. As a consequence, mathematics drifted into the background for long periods. His extensive geodesic work in northern Germany was particularly time consuming. However, one should not overlook that this involvement with practical questions stimulated some of Gauss's mathematical research. An example is differential geometry, where Gauss concerned himself with the question of mapping of a curved surface onto a plane. His geodesical observations led him to the development of many numerical techniques which helped him to master the immense experimental material. Gauss estimated that he worked through more than one million calculations in the course of his life. Nevertheless, one can assume that it would have been better for mathematics had Gauss been able to devote to it his full energies.

We have already seen that Gauss's life was without important external upsetting changes. He spent over 50 years in Göttingen and practically never left the city during his last few decades. He lived simply and modestly, but accumulated a relatively substantial fortune in this way, perhaps a reaction to his poor background.

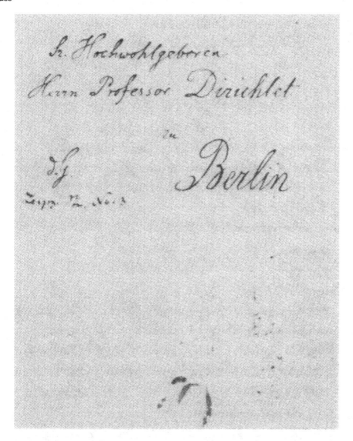

Letter from Carl Friedrich Gauss to Gustav Peter Lejeune Dirichlet (see p. 178 for English translation). Published by permission, Niedersächsische Staats-und Universitätsbibliothek Göttingen.

Although his life was simple, his personal and domestic arrangements, as well as his relations to other scientists, were often complicated and not always easy. Here, we are mainly interested in the latter. We have already noted that it is no exaggeration to say that Gauss did not take much notice of the work of other mathematicians. Jacobi complained that Gauss did not quote any of his or Dirichlet's papers over 20 years. He did not pay any attention to Abel while he was alive; only after Abel's early death did Gauss ask his friend Olbers to see whether he could get his picture for him. Towards the French mathematicians, perhaps also for political reasons, Gauss was lukewarm at best; towards some, he showed considerable animosity. One of the few whom he praised in public was the unhappy Eisenstein, who suffered from illness and depression. In many respects, Eisenstein was the exact opposite of Gauss; his work does not fit at all into Gauss's world of "pauca sed matura."

It seems peculiar that Gauss did not develop many of his most important discoveries though he never tired of claiming his priority over other mathematicians. This created much ill will, and he was often criticized for this. The older he became, the stronger grew his reputation for being inaccessible and unapproachable. He had very few students and avoided contacts whenever he could. To Alexander von Humboldt he appeared to be as icy as a glacier. His unhappy family relations might have contributed to his attitude. As far as we can see, his youth and first period in Göttingen

were free of great aggravations and he seemed to have been quite happy. The four years of his first marriage with Johanna Osthoff, 1805–1809, was a period of serene and satisfied togetherness. Gauss never got over the death of his first wife, which occurred soon after the birth of his second son. He soon remarried, but the second marriage, with Minna Waldeck, was not really happy. Minna was often sick and appears to have had hysterical tendencies. Later, his relationship to his sons from the second marriage was full of tension; eventually, both turned from their father and emigrated to America. Gauss wrote to Wolfgang Bolyai, who had been a friend from the time that both were students in Göttingen: "It is true, my life had much for which the world could envy me. But believe me, the bitter experiences, at least in my life, go very deep; the older one gets, the less one can fight them. They outweigh whatever was good more than a hundred-fold."

References

C. F. Gauss: *Werke*, specifically Vols. I and II (for translations see bibliography).

C. F. Gauss: *Briefwechsel mit Bessel, Gerling, Olbers, Schumacher*. Reprints of the original editions, Georg Olms Verlag, Hildesheim, New York, 1976.

C. F. Gauss: *Mathematisches Tagebuch 1796–1814*, Oswalds Klassiker der exakten Wissenschaften 256, Akad. Verlagsges., Leipzig, 1976.

P. Bachmann: Über Gauss' zahlentheoretische Arbeiten in Materialien für eine wissenschaftliche Biographie von Gauß. Gesammelt von F. Klein und M. Brendel, Teubner, Leipzig, 1911.

G. J. Rieger: Die Zahlentheorie bei C. F. Gauss, in *C. F. Gauss Leben und Werk*. H. Reichardt (ed.), Haude u. Spener, Berlin, 1960.

T. Hall: *Carl Friedrich Gauss*, MIT Press, Cambridge and London, 1970.

H. Wussing: *Carl Friedrich Gauss*, Teubner, Leipzig, 1974.

K. Reich: *Carl Friedrich Gauss 1777/1977*, Heinz Moos Verlag, Munich, 1977.

K. O. May: Gauss, in *Dictionary of Scientific Biography*.

W. K. Bühler: *Gauss (1777–1855), A Biographical Study*. Springer-Verlag, Berlin, Heidelberg, New York, 1981.

O. Neumann: Bemerkungen aus heutiger Sicht über Gauss' Beiträge zur Zahlentheorie, Algebra und Funktionentheorie, Geschichte NTM 16, 2, (1979), 22–39.

Z. I. Borevich and I. R. Shafarevich: *Number Theory*. Transl. by N. Greenleaf, Academic Press, New York, 1966.

CHAPTER 7

Fourier

Jean Baptiste Joseph Fourier (1768–1830) was not a number theorist. He would probably not even have called himself a mathematician, but a physicist. His main area of research was the mathematical theory of heat. He wrote several papers about the topic and one basic book, *Théorie analytique de la chaleur* (Paris, 1822; an English translation was published in 1878). Fourier was a professional politician; as prefect of the Départment d'Isère (at Grenoble), he was closely associated with Napoleon. He accompanied Napoleon on his campaign in Egypt and had the reputation of being quite knowledgeable about that country.

In the preface of his book on heat, Fourier has given a very clear and balanced opinion about the tasks of mathematics and the sciences. Since his convictions coincide with those of so many mathematicians and physicists, we will quote a lengthy passage from this preface. It begins as follows:

> Primary causes are unknown to us; but are subject to simple and constant laws, which may be discovered by observation, the study of them being the object of natural philosophy . . . the most diverse phenomena are subject to a small number of fundamental laws which are reproduced in all the acts of nature. It is recognized that the same principles regulate all the movements of the stars, their form, the inequalities of their courses, the equilibrium and the oscillations of the seas, the harmonic vibrations of air and sonorous bodies, the transmission of light, capillary actions, the undulations of fluids, in fine the most complex effects of all the natural forces, and thus has the thought of Newton been confirmed: quod tam paucis tam multa praestet geometria gloriatur.

Subsequently, Fourier talks about his proper subject, the theory of heat;

and then his statements become more basic again:

> Such are the chief problems which I have solved, and which have never yet been submitted to calculation.

> The principles of the theory are derived, as are those of rational mechanics, from a very small number of primary facts, the causes of which are not considered by geometers, but which they admit as the results of common observations confirmed by all experiment.
> The differential equations of the propagation of heat express the most general conditions, and reduce the physical questions to problems of pure analysis, and this is the proper object of theory.

> The coefficients which they contain are subject to variations whose exact measure is not yet known . . .

Fourier finally develops his basic idea of the simple, immutable, and general laws, recognizable by observation and mathematical description with magnificent and moving intensity and clarity:

> The equations of the movement of heat, like those which express the vibrations of sonorous bodies, or the ultimate oscillations of liquids, belong to one of the most recently discovered branches of analysis, which is very important to perfect. After having established these differential equations their integrals must be obtained; this process consists in passing from a common expression to a particular solution subject to all the given conditions. This difficult investigation requires a special analysis founded on new theorems, whose object we could not in this place make known. The method which is derived from them leaves nothing vague and indeterminate in the solutions, it leads them up to the final numerical applications, a necessary condition of every investigation, without which we should only arrive at useless transformations.
> The same theorems which have made known to us the equations of the movement of heat, apply directly to certain problems of general analysis and dynamics whose solution has for a long time been desired.
> Profound study of nature is the most fertile source of mathematical discoveries. Not only has this study, in offering a determinate object to investigation, the advantage of excluding vague questions and calculations without issue; it is besides a sure method of forming analysis itself, and of discovering the elements which it concerns us to know, and which natural science ought always to preserve: these are the fundamental elements which are reproduced in all natural effects.
> We see, for example, that the same expression whose abstract properties geometers had considered, and which in this respect belongs to general analysis, represents as well the motion of light in the atmosphere, as it determines the laws of diffusion of heat in solid matter, and enters into all the chief problems of the theory of probability.
> The analytical equations, unknown to the ancient geometers, which Descartes was the first to introduce into the study of curves and surfaces, are not restricted to the properties of figures, and to those properties which are the object of rational mechanics; they extend to all general phenomena. There cannot be a language more universal and more simple, more free from errors and from obscurities, that is to say more worthy to express the invariable relations of natural things.
> Considered from this point of view, mathematical analysis is as extensive as nature itself; it defines all perceptible relations, measures times, spaces,

forces, temperatures; this difficult science is formed slowly, but it preserves every principle which it has once acquired; it grows and strengthens itself incessantly in the midst of the many variations and errors of the human mind.

Its chief attribute is clearness; it has no marks to express confused notions. It brings together phenomena the most diverse, and discovers the hidden analogies which unite them. If matter escapes us, as that of air and light, by its extreme tenuity, if bodies are placed far from us in the immensity of space, if man wishes to know the aspect of the heavens at successive epochs separated by a great number of centuries, if the actions of gravity and of heat are exerted in the interior of the earth at depths which will be always inaccessible, mathematical analysis can yet lay hold of the laws of these phenomena. It makes them present and measurable, and seems to be a faculty of the human mind destined to supplement the shortness of life and the imperfection of the senses; and what is still more remarkable, it follows the same course in the study of all phenomena; it interprets them by the same language, as if to attest the unity and simplicity of the plan of the universe, and to make still more evident that unchangeable order which presides over all natural causes.

The problems of the theory of heat present so many examples of the simple and constant dispositions which spring from the general laws of nature; and if the order which is established in these phenomena could be grasped by our senses, it would produce in us an impression comparable to the sensation of musical sound.

The forms of bodies are infinitely varied; the distribution of the heat which penetrates them seems to be arbitrary and confused; but all the inequalities are rapidly cancelled and disappear as time passes on. The progress of the phenomenon becomes more regular and simpler, remains finally subject to a definite law which is the same in all cases, and which bears no sensible impress of the initial arrangement.

We should perhaps keep in mind that Fourier lived during a time of fundamental political and social revolutions. He boldly and decisively defended people who were persecuted during the postrevolutionary terror. He himself was jailed and persecuted as an alleged partisan of Robespierre. Thus Fourier well knew what he meant when he spoke of the errors and changes of the human mind.

The mathematical theory which is the object of Fourier's book is the theory of heat convection described by

$$\Delta v = \frac{\partial^2 v}{x^2} + \frac{\partial^2 v}{\partial y^2} + \frac{\partial^2 v}{\partial z^2} = k \frac{\partial v}{\partial t} ,$$

where v denotes the distribution of heat in a three-dimensional homogeneous body. To treat and solve this partial differential equation, Fourier makes systematic use of the theory of trigonometric series (or "Fourier series"). These series had already appeared in Euler's and D. Bernoulli's work (about the problem of the vibrating string), but Fourier was the first to develop a systematic theory and to recognize that nearly all periodic functions can be expanded as Fourier series. At first, his methods were rejected by Lagrange and consequently did not find general acceptance. The basic result in the theory of Fourier series was proved by Dirichlet.

(7.1) **Theorem.** *Let $f(x)$ be an integral function with period 2π. Assume that for $x_0 \in \mathbb{R}$ the one-sided limits*

$$f(x_0^+) := \lim_{x \downarrow x_0} f(x),$$

$$f(x_0^-) := \lim_{x \uparrow x_0} f(x)$$

and the one-sided derivatives

$$f'(x_0^+) := \lim_{h \downarrow 0} \frac{1}{h} \left(f(x_0 + h) - f(x_0^+) \right),$$

$$f'(x_0^-) := \lim_{h \uparrow 0} \frac{1}{h} \left(f(x_0 + h) - f(x_0^-) \right)$$

exist. Then

$$\frac{1}{2} \left(f(x_0^+) + f(x_0^-) \right) = \frac{1}{2} a_0 + \sum_{n=1}^{\infty} a_n \cos nx_0 + \sum_{n=1}^{\infty} b_n \sin nx_0$$

with

$$a_n = \frac{1}{\pi} \int_0^{2\pi} f(x)\cos(nx)\, dx,$$

$$b_n = \frac{1}{\pi} \int_0^{2\pi} f(x)\sin(nx)\, dx.$$

In the case that $f(x)$ is continuous in x_0, we have $f(x_0^+) = f(x_0^-)$ and therefore

$$f(x_0) = \frac{a_0}{2} + \sum_{n=1}^{\infty} a_n \cos(nx_0) + b_n \sin(nx_0).$$

Two hundred years ago, mathematicians had great difficulties with functions of this kind. It was generally believed that "proper functions" could be expanded into power series; this explains Lagrange's skepticism. Fourier himself gave several interesting applications of his results to classical theorems of analysis. Some of them we have already seen with different proofs. Here we explain some of his examples:

(7.2) EXAMPLES. Consider the function f with $f(x) = x$ in $(-\pi, \pi]$ (continued to be periodic on $\mathbb{R}$). Then all the a_n are 0 since f is an odd function.

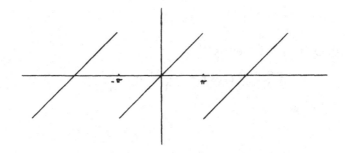

For b_n we have

$$b_n = \frac{1}{\pi} \int_{-\pi}^{\pi} x \sin(nx)\, dx = \frac{1}{\pi} \left[\frac{-x \cos(nx)}{n} + \frac{\sin(nx)}{n^2} \right]_{-\pi}^{\pi}$$

$$= \frac{2}{n} (-1)^{n-1};$$

hence in $(-\pi, \pi]$,

$$x = 2\left(\sin x - \tfrac{1}{2} \sin 2x + \tfrac{1}{3} \sin 3x - + \cdots \right).$$

For $x = \pi/2$ one obtains

$$\frac{\pi}{4} = 1 - \frac{1}{3} + \frac{1}{5} - \frac{1}{7} + - \cdots .$$

Obviously, this is another derivation of Leibniz's series.

As a second example, we consider the continuous function f with $f(x) = |x|$ in $[-\pi, \pi]$ and periodically continued as shown in the figure.

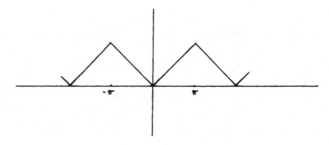

f is an even function; consequently, all the $b_n = 0$. Also,

$$a_0 = \frac{1}{\pi} \int_{-\pi}^{\pi} |x|\, dx = \frac{2}{\pi} \int_0^{\pi} x\, dx = \pi.$$

For reasons of symmetry, one has

$$a_n = \frac{2}{\pi} \int_0^{\pi} x \cos nx\, dx = \frac{2}{\pi} \left[x \frac{\sin nx}{n} + \frac{\cos nx}{n^2} \right]_0^{\pi}$$

$$= \frac{2}{\pi} \left[\frac{\cos n\pi}{n^2} - \frac{1}{n^2} \right]$$

$$= \begin{cases} 0 & \text{for } n \text{ even} \\ -\dfrac{4}{n^2 \pi} & \text{for } n \text{ odd} \end{cases}$$

for $n > 0$. On $[-\pi, \pi]$ one has

$$|x| = \frac{\pi}{2} - \frac{4}{\pi} \left(\cos x + \frac{\cos 3x}{3^2} + \frac{\cos 5x}{5^2} + \cdots \right).$$

For $x = 0$ one gets

$$\frac{\pi^2}{8} = 1 + \frac{1}{3^2} + \frac{1}{5^2} + \cdots .$$

Euler's result

$$\zeta(2) = \sum_{n=1}^{\infty} \frac{1}{n^2} = \frac{\pi^2}{6}$$

follows easily because

$$1 + \frac{1}{2^2} + \frac{1}{3^2} + \frac{1}{4^2} + \cdots$$

$$= \left(1 + \frac{1}{4} + \frac{1}{16} + \cdots\right)\left(1 + \frac{1}{3^2} + \frac{1}{5^2} + \frac{1}{7^2} + \cdots\right)$$

$$= \left(\frac{1}{1 - 1/4}\right)\left(1 + \frac{1}{3^2} + \frac{1}{5^2} + \frac{1}{7^2} + \cdots\right) = \frac{4}{3} \cdot \frac{\pi^2}{8} = \frac{\pi^2}{6}.$$

As a final application, we consider the function f with $f(x) := \cos \alpha x$ on $[-\pi, \pi]$. Here α is a real number, not an integer. Then $b_n = 0$ and

$$a_n = \frac{2}{\pi} \int_0^{\pi} \cos \alpha x \cos nx \, dx$$

$$= (-1)^n \frac{2\alpha \sin \alpha \pi}{\pi(\alpha^2 - n^2)}.$$

Consequently, on $[-\pi, \pi]$,

$$\cos \alpha x = \frac{\sin \alpha \pi}{\pi}\left[\frac{1}{\alpha} - \frac{2\alpha}{\alpha^2 - 1^2}\cos x + \frac{2\alpha}{\alpha^2 - 2^2}\cos 2x - + \cdots\right].$$

If one sets $x = \pi$ and writes x for α, one obtains the expression

$$\pi \cot \pi x = \frac{1}{x} + \frac{2x}{x^2 - 1^2} + \frac{2x}{x^2 - 2^2} + \cdots,$$

i.e., the representation of the cotangent as a partial fraction (cf. page 18). It is valid for every $x \neq 0, \pm 1, \pm 2 \ldots$.

While he was a student in Paris, Dirichlet made the acquaintance of Fourier and was probably the first to understand his theory of trigonometric series. We have already mentioned that he gave a complete proof of (7.1) and successfully applied Fourier's results to number theory. We will discuss this in the next chapter.

Since we have already strayed so far from our main subject, we want to stray even further and mention a connection to the theory of quadratic forms which Fourier or one of his contemporaries might easily have found.

Let us consider the Laplace operator Δ (see page 104) on the space of functions defined on the closed unit cube $W \subset \mathbb{R}^3$, differentiable arbitrarily often and vanishing on the boundary ∂W of W, and the corresponding eigenvalue problem. Obviously,

$$f_{k,l,m}(x, y, z) = \sin(k\pi x)\sin(l\pi y)\sin(m\pi z),$$

where $k, l, m \in \mathbb{Z}$ and $(x, y, z) \in W$ defines a system of linearly indepen-

Jean Baptiste Joseph Fourier

dent eigenfunctions with eigenvalues of the form

$$\lambda_{k,l,m} = -\pi^2(k^2 + l^2 + m^2) =: -\pi^2 n.$$

One can show that these are all the linearly independent eigenfunctions. The multiplicity of the eigenvalue $-\pi^2 n$ is therefore equal to the number of representations of n as sum of three squares minus the number of representations by two squares. (k, l, or m equals 0 if and only if $f = 0$.) The number of representations of a natural number as a sum of three squares can be described only somewhat indirectly. Thus, this observation is not a contribution to mathematical physics, but rather shows in what a simple and direct way seemingly diverse questions are connected.

References

J. Fourier: *Théorie analytique de la chaleur*, Engl. translation, Cambridge, 1878.
J. Ravetz and I. Grattan-Guinness: Fourier, Jean Baptiste Joseph in *Dictionary of Scientific Biography*.
I. Grattan-Guinness: *Joseph Fourier, 1768–1830*, Cambridge, London, 1970.
J. Herivel: *Joseph Fourier, the Man and the Physicist*, Clarendon Press, Oxford, 1975.

CHAPTER 8
Dirichlet

" . . . Dirichlet created a new part of mathematics, the application of those infinite series which Fourier has introduced in the theory of heat to the exploration of the properties of the prime numbers. He has discovered a variety of theorems which . . . are the pillars of new theories." This is what on December 21, 1846 C. G. J. Jacobi wrote in a letter to Alexander von Humboldt. Today, Dirichlet's techniques in number theory are more alive than ever.

In this chapter, we will basically discuss three topics: (1) the still unfinished calculation of the Gaussian sum $G(m)$ (see page 67); (2) the theorem on prime numbers in arithmetical progressions; and (3) the analytical class number formula for a quadratic number field. We will see that these three topics are closely connected.

Let us start with the calculation of the Gaussian sum $G(m)$ (see Dirichlet and Dedekind, *Vorlesungen über Zahlentheorie*, Supplement I). This calculation uses Fourier's result, proved by Dirichlet, about the expansion of periodic functions as trigonometric series (see (7.1)).

Let $m \in \mathbb{N}$, $\epsilon := \exp(2\pi i / m)$,

$$G(m) := \sum_{k=0}^{m-1} \epsilon^{k^2}.$$

As is often done in number theory, set $e(t) := \exp(2\pi i t)$. Clearly, $e(t)$ has period 1 and $G(m)$ can be written as

$$G(m) = \sum_{k=0}^{m-1} e\left(\frac{k^2}{m} \right).$$

Dirichlet then considered the continuously differentiable function

$$f(t) := \sum_{k=0}^{m-1} e\left(\frac{(k+t)^2}{m} \right)$$

for $t \in [0, 1]$, extended it to a periodic function on all of $\mathbb{R}$ (observe that $f(0) = f(1) = G(m)$), and expanded it as a Fourier series. For this it is convenient to use the following equivalent form of (7.1): Every continuous, piecewise continuously differentiable periodic function $f: \mathbb{R} \to \mathbb{C}$ with period 1 can be expanded as a series of the form

$$f(t) = \sum_{n=-\infty}^{+\infty} a_n e(-nt)$$

with

$$a_n = \int_0^1 f(t) e(nt)\, dt.$$

This can easily be derived from (7.1) by making the substitution $x \to 2\pi t$ and separating real and imaginary parts. Applying this to $f(t) = \sum_{k=0}^{m-1} e((k+t)^2/m)$, one obtains

$$f(t) = \sum_{n=-\infty}^{+\infty} a_n e(-nt)$$

with

$$a_n = \int_0^1 \sum_{k=0}^{m-1} e\left(\frac{(k+t)^2}{m} \right) e(nt)\, dt$$

and, specifically,

$$G(m) = f(0) = \sum_{n=-\infty}^{+\infty} a_n.$$

This transforms the finite sum $\sum_{k=1}^{m} e(k^2/m)$ into an infinite series which, as we will see, can be calculated fairly easily. One has

$$a_n = \int_0^1 \sum_{k=0}^{m-1} e\left(\frac{(k+t)^2}{m} \right) e(nt)\, dt$$

$$= \sum_{k=0}^{m-1} \int_0^1 e\left(\frac{(k+t)^2 + mnt}{m} \right) dt$$

$$= \sum_{k=0}^{m-1} \int_0^1 e\left(\frac{(k+t+\frac{1}{2}mn)^2}{m} - \frac{kmn + \frac{1}{4}m^2n^2}{m} \right) dt.$$

Because of $kmn/m \in \mathbb{Z}$ and the periodicity of the exponential function, this expression can be written as

$$= \sum_{k=0}^{m-1} \int_0^1 e\left(\frac{(k + t + \frac{1}{2}mn)^2}{m} \right) e\left(-\frac{1}{4}mn^2 \right) dt$$

$$= e\left(-\frac{1}{4}mn^2 \right) \sum_{k=0}^{m-1} \int_0^1 e\left(\frac{(k + t + mn)^2}{m} \right) dt.$$

The substitution $\tau := k + t + \frac{1}{2}mn$ leads to

$$e\left(-\frac{1}{4}mn^2 \right) \sum_{k=0}^{m-1} \int_{k+(1/2)mn}^{k+1+(1/2)mn} e\left(\frac{\tau^2}{m} \right) d\tau$$

$$= e\left(-\frac{1}{4}mn^2 \right) \int_{(1/2)mn}^{m+(1/2)mn} e\left(\frac{\tau^2}{m} \right) d\tau.$$

Then

$$G(m) = \sum_{n=-\infty}^{+\infty} a_n = \sum_{n=-\infty}^{+\infty} e\left(-\frac{1}{4}mn^2 \right) \int_{(1/2)mn}^{m+(1/2)mn} e\left(\frac{\tau^2}{m} \right) d\tau.$$

For n even, $\frac{1}{4}mn^2$ is an integer and consequently $e(-\frac{1}{4}mn^2) = 1$. For n odd, $n^2 \equiv 1 \bmod 4$ and hence $e(-\frac{1}{4}mn^2) = \eta$ with

$$\eta = \begin{cases} 1 & \text{for} \quad m \equiv 0 \quad \bmod 4, \\ -i & \text{for} \quad m \equiv 1 \quad \bmod 4, \\ -1 & \text{for} \quad m \equiv 2 \quad \bmod 4, \\ i & \text{for} \quad m \equiv 3 \quad \bmod 4, \end{cases}$$

and consequently

$$G(m) = \sum_{n \text{ even}} \int_{(1/2)mn}^{m+(1/2)mn} e\left(\frac{\tau^2}{m} \right) d\tau + \sum_{n \text{ odd}} \eta \int_{(1/2)mn}^{m+(1/2)mn} e\left(\frac{\tau^2}{m} \right) d\tau$$

$$= (1 + \eta) \int_{-\infty}^{\infty} e\left(\frac{\tau^2}{m} \right) d\tau$$

$$= (1 + \eta)\sqrt{m} \int_{-\infty}^{\infty} e(t^2) dt$$

$$= (1 + \eta)\sqrt{m} \left[\int_{-\infty}^{\infty} \cos(2\pi t^2) dt + i \int_{-\infty}^{\infty} \sin(2\pi t^2) dt \right].$$

We calculate the integrals using the following trick. Obviously,

$$G(1) = 1 = (1 - i) \int_{-\infty}^{\infty} e(t^2) dt,$$

hence

$$\int_{-\infty}^{\infty} e(t^2) dt = \frac{1}{1 - i} = \frac{1 + i}{2}.$$

This gives us Euler's result (see page 30)

$$\int_{-\infty}^{\infty} \cos(2\pi t^2)\, dt = \int_{-\infty}^{\infty} \sin(2\pi t^2)\, dt = \tfrac{1}{2}\,.$$

For $G(m)$, we obtain finally:

$$G(m) = \begin{cases} (1 + i)\sqrt{m} & \text{for} \quad m \equiv 0 \mod 4, \\ \sqrt{m} & \text{for} \quad m \equiv 1 \mod 4, \\ 0 & \text{for} \quad m \equiv 2 \mod 4, \\ i\sqrt{m} & \text{for} \quad m \equiv 3 \mod 4, \end{cases}$$

i.e., (6.5).

We now come to Dirichlet's theorem on primes in arithmetic progressions. This theorem is one of the most famous and important theorems in number theory.

Dirichlet starts out with the following question. Are there any prime numbers among the elements of the "arithmetical sequence" ("arithmetical progression")

$$a, a + m, a + 2m, \ldots, a + km, \ldots$$

with $a, m \in \mathbb{N}$, $a < m$, g.c.d.$(a, m) = 1$? If so, are there infinitely many, and in what way are the prime numbers distributed over the sets

$$P_a := \{\, p \text{ prime number} \mid p \equiv a \bmod m \,\},$$

are they perhaps "equidistributed"? All these questions can be answered as follows:

(8.1) There are prime numbers in P_a.
(8.2) There are infinitely many prime numbers in P_a.
(8.3) The $\phi(m)$ disjoint sets P_a contain "asymptotically equally many" prime numbers (ϕ denotes the Euler totient function).

These statements look tantalizingly simple but no simple proof is known. Legendre yielded to the temptation of basing his "proof" of the law of quadratic reciprocity on the unproven statement (8.1) (see page 63). Incidentally, each known proof of (8.1) uses (8.2); (8.2) is not significantly easier to prove than (8.3).

Let P be the set of prime numbers. By (3.11) the series $\sum_{p \in P} 1/p$ diverges. More precisely, one can show that

$$\lim_{s \downarrow 1} \left(\frac{\sum_{p \in P} p^{-s}}{\log(1/(s - 1))} \right) = 1.$$

for real s. This specifically implies the existence of infinitely many prime numbers. (8.3) is proved analogously, by showing that the series $\sum_{p \in P_a} 1/p$ diverges—more precisely, that

$$\lim_{s \downarrow 1} \left(\frac{\sum_{p \in P_a} p^{-s}}{\log(1/(s-1))} \right) = \frac{1}{\phi(m)} .$$

This statement is the precise version of what is meant by (8.3).

One of the difficulties of the proof lies in the fact that one has to "isolate" the prime numbers in the residue classes modulo m. Dirichlet overcame this difficulty by using an idea that was completely new to his contemporaries. He considered (in, of course, a different "language") the so-called characters of the multiplicative group $(\mathbb{Z}/m\mathbb{Z})^*$ of residues mod m which are relatively prime to m. Exploiting elementary properties of these characters gave the required "isolation."

By definition, a character of a finite abelian group G is a homomorphism χ of G into $\mathbb{C}^*$. The characters form an abelian group $\mathrm{Hom}(G, \mathbb{C}^*) =: \hat{G}$, with respect to pointwise multiplication. One can show that for any subgroup H of G the exact sequence $1 \to H \xrightarrow{i} G \xrightarrow{\pi} G/H \to 1$ (with the obvious mappings i and π) induces an exact sequence

$$1 \longrightarrow \widehat{G/H} \xrightarrow{\cdot \pi} \hat{G} \xrightarrow{\cdot i} \hat{H} \longrightarrow 1. \tag{I}$$

G has the same order as $\hat{G}$; one can even show that G and $\hat{G}$ are isomorphic (but they are not canonically isomorphic). The map $G \ni x \mapsto \mathbf{x} \in \hat{\hat{G}}$, $\mathbf{x}(\chi) := \chi(x)$, is a canonical isomorphism $G \cong \hat{\hat{G}}$. Using

$$\sum_{x \in G} \chi(x) = \begin{cases} |G| & \text{if } \chi = 1_G = \text{principal (identity) character of } G, \text{ i.e.,} \\ & \quad \chi(x) = 1 \text{ for all } x \in G \\ 0 & \text{if } \chi \neq 1_G \end{cases} \tag{II}$$

one obtains the relation

$$\sum_{x \in \hat{G}} \chi(x) = \begin{cases} |G| & \text{if } x = 1, \\ 0 & \text{if } x \neq 1. \end{cases} \tag{III}$$

(II) and (III) are called "orthogonality relations." For a proof of these purely algebraic facts, we refer to J. P. Serre, *A Course in Arithmetic*, page 61 ff. In the special case $G = (\mathbb{Z}/m\mathbb{Z})^*$, (III) leads to isolating the prime numbers in the residue classes modulo m. Before making the necessary calculations, we mention that we can interpret a character χ' of $(\mathbb{Z}/m\mathbb{Z})^*$ as a function χ on $\mathbb{Z}$ by setting

$$\chi(a) := \begin{cases} \chi'(a + m\mathbb{Z}) & \text{if } \mathrm{g.c.d}(a, m) = 1, \\ 0 & \text{otherwise.} \end{cases}$$

We then speak of a character modulo m. For example, the only nontrivial character modulo 4 is given by

$$\chi(a) = \begin{cases} 1 & \text{if} \quad a \equiv 1 \quad \mod 4, \\ -1 & \text{if} \quad a \equiv 3 \quad \mod 4, \\ 0 & \text{if} \quad a \equiv 0 \quad \mod 2, \end{cases}$$

and a nontrivial character mod p of order 2 (p a prime number $\neq 2$) is given by

$$\chi_p(a) := \begin{cases} \left(\dfrac{a}{p}\right) & \text{if } p \text{ is not a divisor of } a \\ 0 & \text{otherwise.} \end{cases}$$

To prove (8.3) we must show that

$$\lim_{s \downarrow 1} \frac{\sum_{p \in P_a}(1/p^s)}{\log(1/(s-1))} = \frac{1}{\phi(m)} .$$

Let us first try to understand this statement and transform it so as to see what it really means. In doing this, we do not concern ourselves with questions of convergence. We should, however, mention that this series will certainly converge for real $s > 1$. Therefore, our manipulations are permissible in this case.

Since $|(\mathbb{Z}/m\mathbb{Z})^*| = \phi(m)$, one obtains by (III)

$$\sum_{p \in P_a} \frac{1}{p^s} = \frac{1}{\phi(m)} \sum_{\chi} \chi(a^{-1}) \left(\sum_{p \in P} \frac{\chi(p)}{p^s} \right)$$

$$= \frac{1}{\phi(m)} \left(\sum_{p \nmid m} \frac{1}{p^s} + \sum_{\chi \neq 1} \chi(a^{-1}) \left(\sum_{p \in P} \frac{\chi(p)}{p^s} \right) \right)$$

$$= \frac{1}{\phi(m)} \left(f_1(s) + \sum_{\chi \neq 1} \chi(a^{-1}) f_\chi(s) \right).$$

In this formula, 1 is the principal character modulo m and

$$f_\chi(s) := \sum_{p \in P} \frac{\chi(p)}{p^s} .$$

Since

$$\lim_{s \downarrow 1} \frac{f_1(s)}{\log(1/(s-1))} = 1,$$

we must show that

$$\lim_{s \downarrow 1} \frac{f_\chi(s)}{\log(1/(s-1))} = 0$$

for $\chi \neq 1$. This is best done by showing that $f_\chi(s)$, $\chi \neq 1$, is bounded for

$s \downarrow 1$. To do this, Dirichlet considered, for a character modulo m, the series

$$L(s,\chi) := \sum_{n=1}^{\infty} \frac{\chi(n)}{n^s}.$$

These are now called "Dirichlet L-series." We encountered special examples of such series in Chapter 6. Since $\zeta(s)$ majorizes $|\sum_{n=1}^{\infty} \chi(n)/n^s|$ for $s > 1$ (because $|\chi(n)| \leqslant 1$), this series converges absolutely for $s > 1$. We denote by $L(s,\chi)$ the function that is represented this way for $s > 1$. Because of the multiplicativity of the character χ the L-series can be written as a product for $s > 1$,

$$L(s,\chi) = \prod_{p \in P} \frac{1}{1 - \chi(p)/p^s}.$$

The proof is analogous to the proof of the corresponding formula for the ζ-function: One replaces $(1 - \chi(p)/p^s)^{-1}$ by the geometrical series $\sum_{k=0}^{\infty} (\chi(p)p^{-s})^k$, computes the product of these series, uses the multiplicativity of χ, and rearranges the series. This is permitted because of absolute convergence. We now try to isolate the prime numbers that belong to a certain prime residue class modulo m in this product representation. By taking the logarithm of the product we obtain

$$\log(L(s,\chi)) = \sum_{p \in P} \log\left(\frac{1}{1 - \chi(p)/p^s}\right);$$

χ is complex in general and the logarithm not unique. We determine it uniquely by the series

$$\log \frac{1}{1-x} = x + \frac{1}{2}x^2 + \frac{1}{3}x^3 + \cdots \qquad \text{for} \quad |x| < 1$$

with $x = \chi(p)p^{-s}$. (We justify the process of taking logarithms term by term by first doing it for a finite part of the product,

$$\prod_{p \leqslant t} \frac{1}{1 - \chi(p)/p^s},$$

and then passing to the limit as $t \to \infty$.) This leads to

$$\log(L(s,\chi)) = \sum_{p \in P} \sum_{k=1}^{\infty} \frac{\chi(p)^k}{kp^{ks}}$$

$$= \sum_{p \in P} \frac{\chi(p)}{p^s} + \sum_{p \in P} \sum_{k=2}^{\infty} \frac{\chi(p)^k}{kp^{ks}}$$

$$= f_\chi(s) + F_\chi(s).$$

By the calculation on page 25 and because $|\chi(p)| \leqslant 1$, it follows that $F_\chi(s)$ is bounded. If one wants to show that $f_\chi(s)$, $\chi \neq 1$, is bounded as $s \downarrow 1$, it suffices to show that $\log(L(s,\chi))$, $\chi \neq 1$, is bounded as $s \downarrow 1$. This is the case if the following theorem holds.

(8.4) Theorem. *If* $\chi \neq 1$, *the L-series* $L(s,\chi)$ *converges for* $s{\downarrow}1$ *to a limit* $L(1,\chi) = a \neq 0$.

The proof of convergence is the simpler part of (8.4); the core of Dirichlet's theorem (8.3) is that the L-series does not vanish at 1 for nonprincipal characters χ. There are several ways of proving this. The most direct method is to compute $L(1,\chi)$ for $\chi \neq 1$ directly, and Dirichlet did just this. Later, he realized that $L(1,\chi)$ coincides (up to constant factors) with the logarithm of the fundamental unit of $\mathbb{Q}(\sqrt{p})$ for $m = p$, p a prime number $\equiv 1 \bmod 4$. This result must have left him considerably surprised until he found the general connection between $L(1,\chi)$ and quadratic number fields. The most interesting case is χ nontrivial and real, that is $\chi = \bar{\chi} \neq 1$. The proof of the fact that $L(1,\chi)$ does not vanish can easily be reduced to this case. Dirichlet discovered that there is an integer D which is closely related to m such that $L(1,\chi)$ is in a natural way a factor of the class number of the quadratic number field $\mathbb{Q}(\sqrt{D})$. This is, of course, an exceptionally valuable result because, in addition to proving the nonvanishing of $L(1,\chi)$, it connects an analytic object with a purely algebraic one. We will take this up later, but let us first look at a proof of E. Landau, that $L(1,\chi) \neq 0$, which uses techniques from function theory and is much shorter and easier.

We consider "Dirichlet series," i.e., series of the form

$$\sum_{n=1}^{\infty} \frac{a_n}{n^s}, \qquad a_n \in \mathbb{C},$$

where s not necessarily real. Naturally, $n^s = e^{s \log n}$. The Riemann Zeta-function

$$\zeta(s) := \sum_{n=1}^{\infty} \frac{1}{n^s}$$

is an example. Since $|1/n^s|$ depends only on $\mathrm{Re}(s)$ the series converges absolutely in the half-plane $\mathrm{Re}(s) > 1$. It is possible to continue $\zeta(s)$ to a meromorphic function (with a pole at $s = 1$ having residue 1) in the half-plane $\mathrm{Re}(s) > 0$.

(8.5) Corollary. *There is a holomorphic function* $\psi(s)$, *defined for* $\mathrm{Re}(s) > 0$, *with*

$$\zeta(s) = \frac{1}{s-1} + \psi(s) \qquad \text{for} \quad \mathrm{Re}(s) > 1.$$

PROOF. We write the fairly innocuous function $1/(s-1)$ in a bit more complicated way:

$$\frac{1}{s-1} = \int_1^\infty t^{-s}\, dt = \sum_{n=1}^{\infty} \int_n^{n+1} t^{-s}\, dt.$$

Then

$$\zeta(s) = \frac{1}{s-1} + \sum_{n=1}^{\infty} \left(\frac{1}{n^s} - \int_n^{n+1} t^{-s}\, dt \right)$$

$$= \frac{1}{s-1} + \sum_{n=1}^{\infty} \int_n^{n+1} (n^{-s} - t^{-s})\, dt.$$

We now set

$$\psi_n(s) := \int_n^{n+1} (n^{-s} - t^{-s})\, dt, \qquad \psi(s) := \sum_{n=1}^{\infty} \psi_n(s).$$

Then we have to show that $\psi(s)$ is defined and holomorphic in the half-plane $\mathrm{Re}(s) > 0$. Obviously, each $\psi_n(s)$ has these properties. It is well known that it suffices to show that the series $\sum \psi_n(s)$ converges uniformly on every compact subset of the half-plane $\{s \mid \mathrm{Re}(s) > 0\}$. We know that

$$|\psi_n(s)| \leq \sup_{n \leq t \leq n+1} |n^{-s} - t^{-s}|$$

$$\leq \sup (\text{derivative})$$

$$= \sup \left| \frac{s}{t^{s+1}} \right|$$

$$= \frac{|s|}{n^{\mathrm{Re}(s)+1}}.$$

For any compact set $K \subset \{s \mid \mathrm{Re}(s) > 0\}$ there are $\epsilon > 0$, $c > 0$ such that $\mathrm{Re}(s) > \epsilon$, $|s| < C$; hence $|\psi_n(s)| \leq C/n^{\epsilon+1}$ for $s \in K$. Since the series $\sum_{n=1}^{\infty} C/n^{\epsilon+1}$ converges this proves our proposition.

Although we will not make use of this, we mention that $\zeta(s)$ can be continued to a meromorphic function on $\mathbb{C}$ with a simple pole at $s = 1$ which satisfies the so-called functional equation $\xi(s) = \xi(1-s)$ with $\xi(s) := \pi^{-s/2} \Gamma(s/2) \zeta(s)$ (Γ = the Gamma-function). In the half-plane $\mathrm{Re}(s) < 0$, $\zeta(s)$ only vanishes for $s = -2, -4, -6, \ldots$. These zeros are simple and called the trivial zeros. All further zeros are located inside the strip $0 \leq \mathrm{Re}(s) \leq 1$. Riemann's famous conjecture states that all the zeros lie on the line $\mathrm{Re}(s) = \frac{1}{2}$.

The following theorem summarizes the convergence behavior of a general Dirichlet series, $\sum_{n=1}^{\infty} a_n n^{-s}$, $a_n \in \mathbb{C}$, $s \in \mathbb{C}$. For a proof of this purely function-theoretic theorem we refer to J. P. Serre, *A Course in Arithmetic*, Chapter 6, §2.

(8.6) Theorem. (1) *If the series $\sum a_n n^{-s}$ converges for s_0, it converges for every s with $\mathrm{Re}(s) > \mathrm{Re}(s_0)$. There is a minimal $\rho \in \mathbb{R}$ ($\pm \infty$ are allowed) such that the series converges for $\mathrm{Re}(s) > \rho$. ρ is called the abscissa of convergence of the series.*

We visualize this situation in the figure below. The shaded half-plane, without the line $\mathrm{Re}(s) = \mathrm{Re}(s_0)$, is the area of convergence.

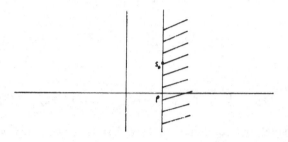

(2) *If the series converges for s_0 it converges uniformly in every sector $\{s \in \mathbb{C} \mid \mathrm{Re}(s - s_0) > 0, \arg(s - s_0) \leqslant \alpha, \alpha < \pi/2\}$. Specifically, the function defined by $\sum a_n n^{-s}$ for $\mathrm{Re}(s) > \mathrm{Re}(s_0)$ is holomorphic by Weierstrass' well-known convergence theorem.*

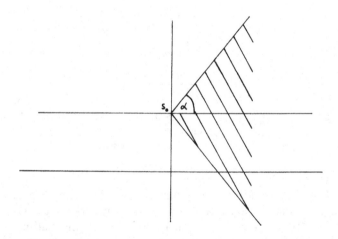

(3) *If there is $C \in \mathbb{R}$, $C > 0$, with $|a_n| < C$, the series converges for $\mathrm{Re}(s) > 1$. The convergence is absolute.*

We can sharpen this result:

(4) *If the partial sums $\sum_{n=1}^{v} a_n$ are bounded, $|\sum_{n=1}^{v} a_n| \leqslant C$, $C \in \mathbb{R}$, then the series converges for $\mathrm{Re}(s) > 0$.*

The following theorem of Landau for Dirichlet series is analogous to the theorem that at least one singular point is located on the circle of convergence of a power series.

(5) *Suppose all the coefficients a_n are real and non-negative. Then the domain of convergence of the series $\sum a_n n^{-s}$ is limited by a singularity (which is located on the real axis) of the function f represented by this series. In other words, if $\sum a_n n^{-s}$ is convergent in the half-plane $\mathrm{Re}(s) > a$, $a \in \mathbb{R}$ and if f can be continued to be holomorphic in a neighborhood of a, then there is $\epsilon > 0$ such that the series converges in $\mathrm{Re}(s) > a - \epsilon$ as well.*

The following statement is analogous to the uniqueness theorem for power series.

(6) *If two Dirichlet series $\sum a_n n^{-s}$, $\sum b_n n^{-s}$ converge in a half-plane and represent the same function, then $a_n = b_n$ for all $n \geqslant 1$.*

Though there are many analogies between power series $\sum c_n z^n$ and Dirichlet series, there is one basic difference. $\sum |c_n| z^n$ and $\sum c_n z^n$ have the same radii of convergence, but this is generally not the case for the abscissas of convergence. $1 - 1/3^s + 1/5^s - 1/7^s + \cdots$ provides a simple example. The abscissa of convergence ρ^+ of the positive series $\sum |a_n| n^{-s}$ is 1, but the abscissa of convergence of $\sum a_n n^{-s}$ is 0. In general, one can show that $\rho^+ - \rho \leqslant 1$.

Let us now apply these general propositions to the specific Dirichlet series

$$L(s, \chi) = \sum_{n=1}^{\infty} \chi(n) n^{-s}$$

with χ a character modulo m. Specifically, we will show that $L(1, \chi) \neq 0$ for $\chi \neq 1$. First we note that $L(s, 1)$ can be continued to a meromorphic function on the half-plane $\text{Re}(s) > 0$ with a simple pole at $s = 1$. This follows from the fact that the function $\zeta(s)$ has these properties by (8.5) and

$$L(s, 1) = \left(\prod_{p \mid m} (1 - p^{-s}) \right) \zeta(s).$$

For $\chi \neq 1$, the series $L(s, \chi)$ converges (converges absolutely) in the half-plane $\text{Re}(s) > 0$ ($\text{Re}(s) > 1$) and, as already stated,

$$L(s, \chi) = \prod_{p \in P} \frac{1}{1 - \chi(p) p^{-s}}$$

for $\text{Re}(s) > 1$. For by (8.6), (4), it suffices to show that the partial sums $\sum_{n=1}^{v} \chi(n)$ are bounded. Since $\chi \neq 1$,

$$\sum_{n=1}^{l+m-1} \chi(n) = \sum_{n \in (\mathbb{Z}/m\mathbb{Z})^*} \chi(n) = 0,$$

by the orthogonality relations (II) on page 113. Hence it suffices to estimate the partial sums $\sum_1^v \chi(n)$ for $v - 1 < m$ which is easily seen to be $\leqslant \phi(m)$.

Specifically, $L(1, \chi)$ is finite for $\chi \neq 1$. It remains to prove that $L(1, \chi) \neq 0$ for $\chi \neq 1$. To do this we consider the product

$$\zeta_m(s) := \prod_{\chi} L(s, \chi)$$

of all the L-series $L(s, \chi)$, where χ runs through the various characters modulo m. If $L(1, \chi) = 0$ for $\chi \neq 1$, $\zeta_m(s)$ would be holomorphic at $s = 1$, for the simple pole of $L(s, 1)$ will be removed at $s = 1$ by this zero. Then, according to the two previous statements, $\zeta_m(s)$ is holomorphic for $\text{Re}(s) > 0$. We show that this is a contradiction by analyzing $\zeta_m(s) = \prod_p \prod_{\chi} 1/(1 - \chi(p) p^{-s})$. Let $\bar{p}$ be the image of p in $(\mathbb{Z}/m\mathbb{Z})^*$ with $p \neq m$;

let $f(p)$ be the order of $\bar{p}$. By definition, $f(p)$ is the smallest natural number $f > 1$ such that $p^f \equiv 1 \mod m$. $f(p)$ divides $\phi(m)$. We set $g(p) := \phi(m)/f(p)$. Using this notation, one has

$$\prod_{\chi} (1 - \chi(p)T) = \left[1 - T^{f(p)}\right]^{g(p)}$$

in the polynomial ring $\mathbb{C}[T]$ where the product extends over all the characters χ of $(\mathbb{Z}/m\mathbb{Z})^*$ because

$$\prod_{w} (1 - wT) = 1 - T^{f(p)}$$

as w runs through all the $f(p)$-th, the roots of unity; there are $g(p)$ characters χ of $(\mathbb{Z}/m\mathbb{Z})^*$ such that $\chi(\bar{p}) = w$ (this can, for instance, be derived from (I)). Then

$$\zeta_m(s) = \prod_{p \nmid m} \frac{1}{\left(1 - p^{-f(p)s}\right)^{g(p)}} .$$

Using standard techniques (geometrical series for the individual factors, etc.) one sees that $\zeta_m(s)$ is a Dirichlet series with real non-negative coefficients which converges in the half-plane $\mathrm{Re}(s) > 1$. Now we show that this series does not converge everywhere for $\mathrm{Re}(s) > 0$. One sees this from the estimate

$$\zeta_m(s_0) \geqq \prod_{p \nmid m} \frac{1}{1 - p^{-f(p)s_0}} \geqq \prod_{p \nmid m} \frac{1}{1 - p^{-\phi(m)s_0}} = L(\phi(m)s_0, 1)$$

(which holds for real s_0) and the fact, noted above, that $L(\phi(m)s_0, 1)$ diverges for $s_0 = 1/\phi(m)$.

So we have seen that if $L(1, \chi) = 0$ for $\chi \neq 1$, the Dirichlet series for $\zeta_m(s)$ with real non-negative coefficients is holomorphic in the half-plane $\mathrm{Re}(s) > 0$; on the other hand, it is not convergent everywhere in this half-plane, a contradiction to (8.6), (5).

This function-theoretic proof is beautiful, but we will see that Dirichlet's direct proof for the nonvanishing of $L(1, \chi)$, $\chi \neq 1$, was more important for the development of number theory. It is contained in his paper "Beweis des Satzes, dass jede unbegrenzte arithmetische Progression, deren erstes Glied und Differenz ganze Zahlen ohne gemeinschaftlichen Faktor sind, unendlich viele Primzahlen enthält." The calculations are not difficult but quite lengthy, so we first summarize the proof. The first step is a derivation of the formula

$$L(1, \chi) = -c_1 \sum_{k=1}^{m-1} \overline{\chi(k)} \log\left(\sin\left(\pi \frac{k}{m}\right)\right) - \frac{i\pi}{m} c_1 \sum_{k=1}^{m-1} \overline{\chi(k)} k$$

with

$$c_1 = \frac{1}{m} \sum_{j=1}^{m} \chi(j)\epsilon^{-j}$$

for a nontrivial character χ modulo m. We show that it suffices to prove the nonvanishing of $L(1,\chi)$ for a real character χ modulo m; without proving it we mention that one can confine oneself to the case when $m = p$ is a prime number. In the next step, we make use of the above formula for the two cases $m = p \equiv 1 \bmod 4$ and $m = p \equiv 3 \bmod 4$. One obtains

$$
c_1 = \begin{cases} \dfrac{1}{\sqrt{p}} & \text{for} \quad p \equiv 1 \quad \bmod 4, \\[3mm] -\dfrac{i}{\sqrt{p}} & \text{for} \quad p \equiv 3 \quad \bmod 4, \end{cases}
$$

and hence, for real χ (i.e., $\chi(k) = (k/p)$),

$$
L(1,\chi) = \begin{cases} -\dfrac{1}{\sqrt{p}} \sum \left(\dfrac{k}{p} \right) \log \sin \left(\pi \dfrac{k}{p} \right) & \text{for} \quad p \equiv 1 \quad \bmod 4, \\[4mm] -\dfrac{\pi}{p\sqrt{p}} \sum_{k=1}^{p-1} \left(\dfrac{k}{p} \right) k & \text{for} \quad p \equiv 3 \quad \bmod 4. \end{cases}
$$

When $p \equiv 3 \bmod 4$ it is easy to see that

$$
L(1,\chi) = -\frac{\pi}{p\sqrt{p}} \left(\sum b - \sum a \right) \neq 0,
$$

where b runs through all quadratic nonresidues and a through all quadratic residues modulo p. This gives us a proof of Dirichlet's theorem for this case. The case $p \equiv 1 \bmod 4$ is more difficult. One finds

$$
L(1,\chi) = \frac{1}{\sqrt{p}} \log \frac{\prod \sin(\pi b/p)}{\prod \sin(\pi a/p)} \qquad (b \text{ and } a \text{ as above})
$$

$$
= \frac{2 \log \eta}{\sqrt{p}} \neq 0
$$

with $\eta \neq 1$ a unit in A_p. This proves the theorem in the case $p \equiv 1 \bmod 4$, but this is only a by-product of our argument giving the formulas for $L(1,\chi)$. We will see that the last formula leads to a deep connection between $L(1,\chi)$ and $\mathbb{Q}(\sqrt{p})$. Dirichlet investigated this and derived an expression for the class number of an arbitrary quadratic number field which contains $L(1,\chi)$ in an essential way (here, χ is a real character defined modulo the discriminant of the quadratic number field; see page 128–129). We do not have the space to give a complete proof of the so-called analytic class number formula, but we will summarize the essential steps of the analytical proof. The main step consists of deriving an alternate expression for $L(1,\chi)$. We perform this calculation when χ is the character of a quadratic number field with class number 1. It may seem odd that we compute $L(1,\chi)$ only in this case when one wants to derive the class number formula; but this case does indeed contain all the essential elements of the analytical proof. Everything else is basically an algebraic problem; see page 136 ff. The so-called decomposition theorem for A_d (see page 136) yields, for $L(1, \chi)$,

$$L(1, \chi) = \lim_{s \downarrow 1} (s - 1) \zeta_K(s),$$

where $\zeta_K(s)$ is the ζ-function of the quadratic number field $K = \mathbb{Q}(\sqrt{d})$; see page 133. One interprets the representation of this ζ-function by a series as a limit of a Riemann sum for a double integral. The latter can be calculated easily (see page 134). Except when $d = -1, -3$, one obtains

$$L(1, \chi) = \begin{cases} \dfrac{2 \log \epsilon}{\sqrt{D}} & \text{if } d > 0, \\[3mm] \dfrac{\pi}{\sqrt{|D|}} & \text{if } d < 0. \end{cases}$$

D is the discriminant of $\mathbb{Q}(\sqrt{d})$ and ϵ the fundamental unit in $\mathbb{Q}(\sqrt{d})$.

Step I

Let ϵ be a primitive mth root of unity. One considers the series

$$\frac{\epsilon}{1} + \frac{\epsilon^2}{2} + \frac{\epsilon^3}{3} + \cdots .$$

Since the partial sums are bounded ($|\sum_k \epsilon^k| \leq m$), the series converges by (8.6), (4). Then

$$\frac{\epsilon}{1} + \frac{\epsilon^2}{2} + \frac{\epsilon^3}{3} + \cdots = \log \frac{1}{1 - \epsilon},$$

according to Abel's limit theorem. Consequently,

$$\frac{\epsilon^k}{1} + \frac{\epsilon^{2k}}{2} + \frac{\epsilon^{3k}}{3} + \cdots = \log \frac{1}{1 - \epsilon^k}.$$

The linear system

$$\begin{pmatrix} \epsilon & \epsilon^2 & \cdots & \epsilon^m \\ \epsilon^2 & & \cdots & \epsilon^{2m} \\ \vdots & \vdots & & \vdots \\ \epsilon^m & \epsilon^{2m} & \cdots & \epsilon^{m^2} \end{pmatrix} \begin{pmatrix} c_1 \\ \vdots \\ \vdots \\ c_m \end{pmatrix} = \begin{pmatrix} \chi(1) \\ \vdots \\ \vdots \\ \chi(m) \end{pmatrix} \qquad (*)$$

can be solved for $c_1, \ldots, c_m$ because the matrix A is nonsingular. One way to see this is to square it.

$$A^2 = \left(\sum_{k=1}^{m} \epsilon^{ik} \epsilon^{jk} \right)_{i,j} = \left(\sum_{k=1}^{m} \epsilon^{(i+j)k} \right)_{i,j}$$

$$= \begin{bmatrix} 0 & \cdots & 0 & m & 0 \\ \vdots & & & & \vdots \\ 0 & & \ddots & & \\ m & & & & \vdots \\ 0 & \cdots & & & m \end{bmatrix}.$$

Below, we will solve (∗) explicitly. For now, we write $L(1,\chi)$ in the form

$$L(1,\chi) = \frac{\chi(1)}{1} + \frac{\chi(2)}{2} + \cdots + \frac{\chi(m)}{m} + \frac{\chi(1)}{m+1} + \cdots$$

$$= c_1 \log \frac{1}{1-\epsilon} + c_2 \log \frac{1}{1-\epsilon^2} + \cdots + c_{m-1} \log \frac{1}{1-\epsilon^{m-1}},$$

using the solution of (∗):

$$(\chi(1), \ldots, \chi(m)) = \sum_{k=1}^{m} c_k (\epsilon^k, \epsilon^{2k}, \ldots, \epsilon^{mk}), \qquad \sum_{k=1}^{m} c_k = 0.$$

With $\epsilon = \exp(2\pi i / m)$ and, hence, $\epsilon^k = \exp(2\pi i k / m)$, one obtains

$$\log \frac{1}{1-\epsilon^k} = \log \frac{1}{1-\exp(2\pi i k / m)}$$

$$= \log \frac{\exp(-\pi i k / m)}{-\exp(\pi i k / m) + \exp(-\pi i k / m)}$$

$$= \log \frac{\exp(-\pi i k / m)}{-2i \sin(\pi k / m)}.$$

Since $i = e^{i\pi/2}$ one can write this as

$$= \log \frac{\exp(i\pi/2)\exp(-\pi i k / m)}{2 \sin(\pi k / m)}$$

$$= \left(\frac{i\pi}{2} - \frac{i\pi k}{m} \right) - \log 2 - \log \left(\frac{\sin \pi k}{m} \right),$$

and consequently

$$L(1,\chi) = -\sum_{k=1}^{m-1} c_k \left(\frac{\pi i k}{m} + \log \left(\frac{\sin \pi k}{m} \right) \right) + \left(\sum_{k=1}^{m-1} c_k \right) \left(\frac{i\pi}{2} - \log 2 \right).$$

To solve the system (∗) we multiply both sides of (∗) by the coefficient matrix A and obtain, using the above expression for A^2,

$$\begin{bmatrix} 0 & \cdots & 0 & m & 0 \\ \vdots & & \ddots & & \\ 0 & & & \ddots & \\ m & & & & \\ 0 & \cdots & & & m \end{bmatrix} \begin{bmatrix} c_1 \\ \vdots \\ c_m \end{bmatrix} = \begin{bmatrix} \sum \chi(j)\epsilon^j \\ \sum \chi(j)\epsilon^{2j} \\ \vdots \\ \sum \chi(j)\epsilon^{mj} \end{bmatrix}.$$

For the coefficients, one obtains

$$c_m = \frac{1}{m} \sum_j \chi(j)\epsilon^{mj} = 0,$$

$$c_k = \frac{1}{m} \sum_{j=1}^{m} \chi(j)\epsilon^{-kj}.$$

Lemma. $(c_1, c_2, \ldots, c_m) = c_1(\overline{\chi(1)}, \overline{\chi(2)}, \ldots, \overline{\chi(m)})$.

PROOF. We have to show that $c_k = c_1\overline{\chi(k)}$.

First Case. Let us assume that k is relatively prime to m. Then there is a j with $kj \equiv 1 \bmod m$. Then $\chi(k)\chi(j) = 1$, i.e., $\overline{\chi(k)} = \chi(j)$. Hence

$$c_1\overline{\chi(k)} = \frac{1}{m}\left(\chi(1)\epsilon^{-1} + \chi(2)\epsilon^{-2} + \cdots + \chi(m)\epsilon^{-m}\right)\overline{\chi(k)}$$

$$= \frac{1}{m}\left(\chi(j)\epsilon^{-1} + \chi(2j)\epsilon^{-2} + \cdots + \chi(mj)\epsilon^{-m}\right)$$

$$= \frac{1}{m}\left(\chi(1)\epsilon^{-k} + \chi(2)\epsilon^{-2k} + \cdots + \chi(m)\epsilon^{-mk}\right)$$

$$= c_k.$$

Second Case. k is not relatively prime to m. Then one can write $k = pr$, $m = pn$. We have to show that $c_k = 0$ ($\chi(k) = 0!$). We start with

$$c_k = \frac{1}{m}\left(\chi(1)\epsilon^{-pr} + \chi(2)\epsilon^{-2pr} + \cdots + \chi(m)\epsilon^{-mpr}\right).$$

If $ar \equiv br \bmod n$, then $arp \equiv brp \bmod m$, hence $\epsilon^{-arp} = \epsilon^{-brp}$ and

$$c_k = \frac{1}{m}\left(\left(\sum_{t \equiv 1(n)} \chi(t)\right)\epsilon^{-pr} + \cdots + \left(\sum_{t \equiv n-1(n)} \chi(t)\right)\epsilon^{-(n-1)pr}\right).$$

It suffices to show that $\sum_{t \equiv 1(n)}\chi(t) = 0$, $\sum_{t \equiv 2(n)}\chi(t) = 0, \ldots$. This follows from the orthogonality relations. See page 113.

For $L(1,\chi)$ we now have the formula

$$L(1,\chi) = -c_1 \sum_{k=1}^{m-1} \overline{\chi(k)} \log\sin(\pi k/m) - \frac{i\pi}{m}c_1 \sum_{k=1}^{m-1} \overline{\chi(k)}\, k. \qquad (**)$$

Step II

Before continuing with our computation of this expression we claim that it will suffice to show that $L(1,\chi) \neq 0$ for a real character $\chi \neq 1$. This is because of the following lemma.

Lemma. *Let $\chi \neq 1$ be a nonreal character. Then $L(1,\chi) \neq 0$.*

PROOF. Let us assume there is a nonreal character $\chi \neq 1$ with $L(1,\chi) = 0$. If χ is a character, then $\bar{\chi}$, defined by $\bar{\chi}(x) = \overline{\chi(x)}$ is also a character, and $\chi \neq \bar{\chi}$. If χ is real, then

$$L(s,\bar{\chi}) = \sum_{n=1}^{\infty} \overline{\chi(n)}\, n^{-s} = \overline{L(s,\chi)}\,.$$

This means that we have shown that $L(1,\chi) = 0$ implies $L(1,\bar{\chi}) = 0$ which is impossible by the following

Lemma. *If there is a character $\chi_2 \neq 1$ with $L(1,\chi_2) = 0$, then there is no character χ_3 different from χ_2 and 1, such that $L(1,\chi_3) = 0$.*

PROOF. For real $s > 1$ one has

$$\frac{1}{\phi(m)} \sum_{\chi} \overline{\chi(a)} \log L(s,\chi) = \sum_{P} \sum_{n=1}^{\infty} \frac{1}{np^{ns}} \qquad p^n \equiv a \mod m.$$

For $a = 1$, one obtains

$$\frac{1}{\phi(m)} \sum_{\chi} \log L(s,\chi) = \sum_{p \equiv 1(m)} p^{-s} + \frac{1}{2} \sum_{p^2 \equiv 1(m)} p^{-2s} + \frac{1}{3} \sum_{p^3 \equiv 1(m)} p^{-3s} + \cdots,$$

and hence

$$\prod_{\chi} L(s,\chi) \geqslant 1$$

and

$$(s-1)L(s,1)\frac{L(s,\chi_2)L(s,\chi_3)}{(s-1)(s-1)} \prod_{\chi \neq 1,\chi_2,\chi_3} L(s,\chi) \geqq \frac{1}{s-1}. \qquad (***)$$

Let us now assume that

$$L(1,\chi_2) = L(1,\chi_3) = 0.$$

Then

$$\lim_{s \downarrow 1} \frac{L(s,\chi_2) - L(1,\chi_2)}{s-1} = \lim_{s \downarrow 1} \frac{L(s,\chi_2)}{s-1} = L'(1,\chi_2).$$

An analogous expression holds for $L(s,\chi_3)$. Since $\lim_{s \downarrow 1}(s-1)\zeta(s) = 1$ and $L(s,1) = \prod_{p/m}(1 - p^{-s})\zeta(s)$, one has $\lim_{s \downarrow 1}(s-1)L(s,1) = \phi(m)/m$. So it follows that the left-hand side of $(***)$ converges towards a fixed finite limit for $s \downarrow 1$, while the right-hand side diverges.

Step III

We now continue with the computation of $L(1,\chi)$ for a real character χ modulo m, $\chi \neq 1$; we are aiming towards the analytic class number formula.

It is possible to reduce the proof that $L(1,\chi) \neq 0$ to the case that $m = p$ is an odd prime number (see Dirichlet's original paper, paragraph 7). The only nontrivial real character χ modulo p is given by

$$\chi(k) = \left(\frac{k}{p} \right).$$

Then

$$c_1 = \frac{1}{p} \sum_{j=1}^{p-1} \left(\frac{j}{p} \right) \epsilon^{-j}, \qquad \overline{c_1} = \frac{1}{p} \sum_{j=1}^{p-1} \left(\frac{j}{p} \right) \epsilon^{j}.$$

The second expression is a Gaussian sum. By (6.5) we obtain

$$\overline{c_1} = \begin{cases} \dfrac{1}{p} \sqrt{p} & \text{if } p \equiv 1 \mod 4, \\[2mm] \dfrac{1}{p} i\sqrt{p} & \text{if } p \equiv 3 \mod 4, \end{cases}$$

$$c_1 = \begin{cases} \dfrac{1}{\sqrt{p}} & \text{if } p \equiv 1 \mod 4, \\[2mm] -\dfrac{i}{\sqrt{p}} & \text{if } p \equiv 3 \mod 4. \end{cases}$$

Since χ is real, $L(1,\chi)$ is real as well (and obviously ≥ 0 if one looks at the expansion $L(1,\chi) = \prod_p (1 - \chi(p)p^{-1})^{-1}$). Then it follows from (**) that

$$L(1,\chi) = \begin{cases} -\dfrac{1}{\sqrt{p}} \sum_{k=1}^{p-1} \left(\dfrac{k}{p} \right) \log \sin \left(\dfrac{\pi k}{p} \right) & \text{if } p \equiv 1 \mod 4, \\[4mm] -\dfrac{\pi}{p\sqrt{p}} \sum_{k=1}^{p-1} \left(\dfrac{k}{p} \right) k & \text{if } p \equiv 3 \mod 4. \end{cases}$$

When $p \equiv 3 \mod 4$, $L(1,\chi)$ can be written as

$$L(1,\chi) = \frac{\pi}{p\sqrt{p}} \left(\sum b - \sum a \right),$$

where b runs through all k with $(k/p) = -1$ and a through all k with $(k/p) = +1$. If one calculates modulo 2, $\sum b - \sum a$ equals $\sum b + \sum a = \sum_{k=1}^{p-1} k = p(p-1)/2$. This latter number is odd because $p = 4n + 3$. Specifically, $\sum b - \sum a \neq 0$ and consequently $L(1,\chi) \neq 0$.

Let us, for example, look at the case $p = 3$:

$$\chi(1) = 1, \qquad \chi(2) = -1, \qquad \chi(3) = 0,$$

$$L(1,\chi) = 1 - \frac{1}{2} + \frac{1}{4} - \frac{1}{5} + \frac{1}{7} - \frac{1}{8} + - \cdots$$

$$= \frac{\pi}{3\sqrt{3}}(2 - 1) = \frac{\pi}{3\sqrt{3}},$$

Again, this is a result which Euler had already found.

Our considerations contain a very interesting consequence:

(8.7) Theorem. $\sum b - \sum a > 0.$

This means that for $p \equiv 3 \bmod 4$ the quadratic residues are on the average smaller than the nonresidues in the interval 1 to p. Although so simple, this statement is a very deep number-theoretical result; one can only admire Dirichlet's perception since he quite rightly remarked that it would probably be very difficult to prove this result in any other way. In fact, no proof that is essentially different from this one seems to be known.

Step IV

We now investigate the more difficult case $m = p \equiv 1 \bmod 4$. With a, b as above, we have

$$L(1,\chi) = \frac{1}{\sqrt{p}} \log \frac{\prod \sin(\pi b / p)}{\prod \sin(\pi a / p)}.$$

In order to obtain more information about $L(1,\chi)$ from this expression, one shows that

$$\frac{\prod \sin(\pi b / p)}{\prod \sin(\pi a / p)} = \frac{s + t\sqrt{p}}{-s + t\sqrt{p}}$$

with $s + t\sqrt{p} \in A_p$ and $s^2 - t^2 p = \pm 4$. This last expression can thus be written as

$$= \pm \frac{\left(s + t\sqrt{p}\right)^2}{4} = \pm \left(\frac{s + t\sqrt{p}}{2}\right)^2$$

and $\eta := (s + t\sqrt{p})/2$ is a unit in A_p. So we see that the computation of $L(1,\chi)$ leads to an equation of the form $x^2 - dy^2 = \pm 4$.

There is a connection between this equation and the equation $x^2 - dy^2 = \pm 1$ discussed in Chapter 5. Let us assume that $d \equiv 1 \bmod 4$. If $d \equiv 5$

mod 8 and $\epsilon = x + y\sqrt{d}$ with $2x, 2y, x + y \in \mathbb{Z}$ a unit in A_d, then

$$\epsilon^3 = \frac{s^3 + 3st^2d + (3s^2t + t^3d)\sqrt{d}}{8}$$

with $x = s/2$, $y = t/2$. We also know that $s^3 + 3st^2d = s(s^2 + 3t^2d) \equiv 16$ $\equiv 0 \mod 8$ and $3s^2t + t^3d = t(3s^2 + t^2d) \equiv 0 \mod 8$. Hence $\epsilon^3 = u + v\sqrt{d}$ with $u, v \in \mathbb{Z}$, and (u, v) is a solution of $x^2 - dy^2 = \pm 1$. The case $d \equiv 1$ mod 8 is even simpler.

We have seen that the equation involving ± 4 leads to the one involving ± 1. In the case under consideration, a fundamental unit ϵ exists in A_p, i.e., ϵ has the form

$$\epsilon = \frac{s_0 + t_0\sqrt{p}}{2},$$

where $s_0, t_0 > 0$ is a minimal solution of $x^2 - py^2 = \pm 4$. Every other unit is a power of ϵ. Dirichlet was familiar with all this (from Lagrange's work). His great discovery was that

$$\eta = \epsilon^h$$

for the η defined above, where h is the class number of $\mathbb{Q}(\sqrt{p})$. Hence one obtains the following relation for h (or for $L(1, \chi)$):

$$h \frac{2 \log \epsilon}{\sqrt{p}} = L(1, \chi).$$

Specifically, $L(1, \chi) \neq 0$ because $\epsilon \neq 1$, $h \neq 0$. Here we observe explicitly that Dirichlet's proof naturally leads to the problem $\eta = \epsilon^?$ that is answered by the class number formula. Today, this formula is often proved in textbooks without showing how one would be led to it—it miraculously appears out of nothing.

(8.8) **Theorem** (The Analytic Class Number Formula). *Let d be a square-free integer $\neq -1, -3$ (these cases have already been treated in Chapter 6). Then the following formula holds for the class number h of $\mathbb{Q}(\sqrt{d})$.*

$$h = \begin{cases} \dfrac{\sqrt{D}}{2 \log \epsilon} L(1, \chi) & \text{for } d > 0, \\[3ex] \dfrac{\sqrt{|D|}}{\pi} L(1, \chi) & \text{for } d < 0. \end{cases}$$

D is the discriminant of $\mathbb{Q}(\sqrt{d})$ and ϵ the fundamental unit in A_d for $d > 0$. χ

is the following character modulo $|D|$:

$$
\chi(k) = \begin{cases}
\prod_{p/d}\left(\dfrac{k}{p}\right) & \text{for} \quad d \equiv 1 \quad \bmod 4, \\[3mm]
(-1)^{(k-1)/2}\prod_{p/d}\left(\dfrac{k}{p}\right) & \text{for} \quad d \equiv 3 \quad \bmod 4, \\[3mm]
(-1)^{(k^2-1)/8+(k-1)(\delta-1)/4}\prod_{p/\delta}\left(\dfrac{k}{p}\right) & \text{for} \quad d = 2\delta, \quad \delta \text{ odd.}
\end{cases}
$$

The expressions $(-1)^{(k-1)/2}$ and $(-1)^{(k^2-1)/8}$ are defined when $d \equiv 2, 3$ mod 4 since k is odd. As χ has been defined as a product of characters it is itself a character called the (real or quadratic) character belonging to $\mathbb{Q}(\sqrt{d}\,)$.

A complete proof of this formula is too lengthy to fit within the framework of this book, and we will confine ourselves to a few special cases which show the pattern of the general proof. Towards the end of this chapter, we will make a few remarks about what is necessary for a general proof.

Let us first return to the special case $d = p \equiv 1 \bmod 4$. First, we have to prove the relation

$$
\frac{\prod \sin(\pi b/p)}{\prod \sin(\pi a/p)} = \frac{s + t\sqrt{p}}{-s + t\sqrt{p}},
$$

where b runs through all k with $(k/p) = -1$ and a through all k with $(k/p) = 1$. Here we make use of a few facts from algebra. The primitive pth root of unity $\epsilon = \exp(2\pi i/p)$ (p, for now, is an arbitrary prime number) is a zero of the so-called cyclotomic polynomial

$$
\phi_p(x) := \frac{x^p - 1}{x - 1} = x^{p-1} + x^{p-2} + \cdots + 1 \in \mathbb{Q}[x].
$$

ϕ_p is irreducible over $\mathbb{Q}$, which can easily be proved with the help of Eisenstein's criterion after replacing x by $x + 1$. This means that the field extension $\mathbb{Q}(\epsilon)/\mathbb{Q}$ has degree $p - 1$. By (6.3), the relation $S^2 = \pm p$ holds for the Gauss sum $S = \sum_{k=1}^{p-1}(\frac{k}{p})\epsilon^k$. Thus $\mathbb{Q}(\epsilon)$ contains the quadratic field $\mathbb{Q}(\sqrt{\pm p}\,)$, with the positive sign when $p \equiv 1 \bmod 4$ and the negative sign when $p \equiv 3 \bmod 4$. The expressions $\epsilon, \epsilon^2, \ldots, \epsilon^{p-1}$ are the zeros of $\phi_p(x)$ and $\mathbb{Q}(\epsilon) = \mathbb{Q}(\epsilon^k)$ for $k = 1, \ldots, p - 1$. The minimal polynomial of ϵ^k has degree $(p - 1)/2$ over $\mathbb{Q}(\sqrt{\pm p}\,)$; $\phi_p(x)$ decomposes over $\mathbb{Q}(\sqrt{\pm p}\,)$ and there is a factorization

$$
\phi_p(x) = f(x)g(x)
$$

with degree $f =$ degree $g = (p - 1)/2$. We now claim the following:

If, without loss of generality, ϵ is a zero of $f(x)$ then the ϵ^a are the zeros of $f(x)$ and ϵ^b the zeros of $g(x)$.

PROOF. Since the Galois group G of the extension $\mathbb{Q}(\epsilon)/\mathbb{Q}$ is isomorphic to $(\mathbb{Z}/p\mathbb{Z})^*$ we can associate to $r \in (\mathbb{Z}/p\mathbb{Z})^*$ the element $\sigma_r(\epsilon)$ $:= \epsilon^r$. r determines an element of the Galois group H of the extension $\mathbb{Q}(\epsilon)/\mathbb{Q}(\sqrt{\pm p})$ if and only if r is a square in $(\mathbb{Z}/p\mathbb{Z})^*$, i.e., if $(\frac{r}{p}) = 1$. With ϵ, also $\sigma(\epsilon) = \epsilon^r$, $\sigma \in H$, is a zero of f. Thus f has the zeros ϵ^a as claimed and g the remainder of the zeros, namely, ϵ^b.

Since f and g are polynomials with coefficients in $\mathbb{Q}(\sqrt{\pm p})$ we may write

$$f(x) = f_0(x) + f_1(x)\sqrt{\pm p},$$

$$g(x) = g_0(x) + g_1(x)\sqrt{\pm p}$$

with polynomials $f_0, f_1, g_0, g_1 \in \mathbb{Q}[x]$. We claim that

$$f_0 = g_0, \qquad f_1 = -g_1.$$

PROOF. Let σ be a generator of the cyclic Galois group G of $\mathbb{Q}(\epsilon)/\mathbb{Q}$ e.g., $\sigma: \epsilon \to \epsilon^r$, where r is a generating element of $(\mathbb{Z}/p\mathbb{Z})^*$. Then $\sigma(\epsilon^a) = \epsilon^{ar}$, and since ar is a quadratic nonresidue in $(\mathbb{Z}/p\mathbb{Z})^*$, we have

$$(\sigma f)(x) = \prod_a (x - \epsilon^{ar}) = \prod_b (x - \epsilon^b) = g(x).$$

σ restricted to the field $\mathbb{Q}(\sqrt{\pm p})$ gives conjugation, i.e., $\sigma: \alpha + \beta\sqrt{\pm p}$ $\to \alpha - \beta\sqrt{\pm p}$. The statement follows immediately.

This means that the cyclotomic polynomial $\phi_p(x)$ can be decomposed in the form

$$\phi_p(x) = \left(f_0(x) + f_1(x)\sqrt{\pm p} \right)\left(f_0(x) - f_1(x)\sqrt{\pm p} \right)$$

$$= f_0(x)^2 \mp f_1(x)^2 p.$$

Then, for the case $p \equiv 1 \bmod 4$, we claim

$$f(1) = 2^{(p-1)/2} \prod_a \sin\left(\frac{\pi a}{p} \right).$$

PROOF. Clearly,

$$2^{(p-1)/2} \prod_a \sin\left(\frac{\pi a}{p} \right) = 2^{(p-1)/2} \prod_a \left(\frac{1}{2i} \exp\left(\frac{i\pi a}{p} \right) - \exp\left(\frac{-i\pi a}{p} \right) \right)$$

$$= (-i)^{(p-1)/2} \prod_a \exp\left(\frac{-i\pi a}{p} \right) \prod_a \left(\exp\left(\frac{2\pi i a}{p} \right) - 1 \right)$$

$$= (-1)^{(p-1)/4} \prod_a \exp\left(\frac{-i\pi a}{p} \right) \prod_a \left(1 - \exp\left(\frac{2\pi i a}{p} \right) \right).$$

Since $(-1/p) = 1$, one obtains $(a'/p) = 1$ by setting $a' = p - a$. Then

$$\exp\left(\frac{-i\pi a}{p}\right)\exp\left(\frac{-i\pi a'}{p}\right) = \exp(-i\pi) = -1$$

and

$$\prod_a \exp\left(\frac{-i\pi a}{p}\right) = (-1)^{(p-1)/4}$$

and, consequently,

$$(-1)^{(p-1)/4}\prod_a \exp\left(\frac{-i\pi a}{p}\right)\prod_a\left(1 - \exp\left(\frac{2\pi ia}{p}\right)\right)$$

$$= \prod_a\left(1 - \exp\left(\frac{2\pi ia}{p}\right)\right) = f(1).$$

Similarly, one shows

$$g(1) = 2^{(p-1)/2}\prod_b \sin\left(\frac{\pi b}{p}\right).$$

All of our further computations are for the case $p \equiv 1 \bmod 4$. Since it is an element of $\mathbb{Q}(\sqrt{p})$, $g(1)$ can be written as

$$g(1) = k + l\sqrt{p}$$

with $k, l \in \mathbb{Q}$. Then, by what we have shown above,

$$f(1) = k - l\sqrt{p}$$

and

$$\frac{\prod \sin(\pi b/p)}{\prod \sin(\pi a/p)} = \frac{g(1)}{f(1)} = \frac{k + l\sqrt{p}}{k - l\sqrt{p}}.$$

The expression

$$p = \phi_p(1) = f(1)g(1) = k^2 - l^2 p$$

shows that both l and $k \neq 0$. Then, $(k + l\sqrt{p})/(k - l\sqrt{p}) = 1$ and consequently,

$$\log\frac{\prod \sin(\pi b/p)}{\prod \sin(\pi a/p)} \neq 0$$

which means

$$L(1,\chi) \neq 0.$$

We note that we have given a complete proof for the nonvanishing of $L(1,\chi)$ for any prime number $m = p$.

Step V

In deriving the class number formula one is led to the expression

$$\frac{\prod \sin(\pi b/p)}{\prod \sin(\pi a/p)} = \frac{k + l\sqrt{p}}{k - l\sqrt{p}}$$

which has to be investigated.

We show that

$$k + l\sqrt{p} \in A_p,$$

i.e., that $2k, 2l, k + l \in \mathbb{Z}$ for $p \equiv 1 \mod 4$ in this case. Anybody who is partly familiar with the fundamentals of algebraic number theory knows that any product of algebraic integers is again an algebraic integer. This is also true for $g(1)$:

$$k + l\sqrt{p} = g(1) = \prod_b \left(1 - \exp\left(\frac{2\pi i b}{p} \right) \right).$$

Since we are assuming as little as possible in this book, we will give a direct proof. First, one has to show that

$$\mathbb{Z}[\epsilon] = \mathbb{Z} \oplus \mathbb{Z}\epsilon \oplus \cdots \oplus \mathbb{Z}\epsilon^{p-2}$$

is a ring, i.e., closed with respect to multiplication. This one easily sees by using the relation

$$\epsilon^{p-1} = -\epsilon^{p-2} - \epsilon^{p-3} - \cdots - 1.$$

Next, one shows that

$$\mathbb{Q}(\sqrt{p}) \cap \mathbb{Z}[\epsilon] \subset A_p.$$

To see this, we assume that an element $k + l\sqrt{p} \in \mathbb{Q}(\sqrt{p}) \cap \mathbb{Z}[\epsilon]$ is not in A_p. According to our characterization of A_p on page 77, $(k + l\sqrt{p}) + (k - l\sqrt{p}) = 2k \notin \mathbb{Z}$ or $k^2 - pl^2 \notin \mathbb{Z}$. This contradicts $\mathbb{Z}[\epsilon] \cap \mathbb{Q} = \mathbb{Z}$ which is true because $1, \epsilon, \ldots, \epsilon^{p-2}$ are linearly independent.

Since $k, l \in \frac{1}{2}\mathbb{Z}$ and $p = k^2 - l^2 p$, there is one and only one $h \in \frac{1}{2}\mathbb{Z}$ such that $k = hp$ and $1 = ph^2 - l^2$ or

$$l^2 - ph^2 = -1$$

and consequently

$$\frac{k + l\sqrt{p}}{k - l\sqrt{p}} = \frac{l + h\sqrt{p}}{-l + h\sqrt{p}} = \frac{\left(l + h\sqrt{p} \right)^2}{1} = \eta^2,$$

where η is a unit in A_p.

Step VI

To derive the class number formula one has to find yet another way to express $L(1,\chi)$. To illustrate this procedure we look at the case $\mathbb{Q}(\sqrt{d}\,)$, $d \equiv 1 \bmod 4$ having class number 1 and a fundamental unit ϵ of negative norm, for example, $d = 5$. To confine ourselves to the case of class number 1 is certainly a restriction but the general case can be handled similarly. First, let us look at the series

$$\sum_a \frac{1}{|N(a)|^s} \, ,$$

for $K = \mathbb{Q}(\sqrt{d}\,)$ and real $s > 1$. We have the sum over a complete system of representatives of the equivalence classes of associated elements $\neq 0$ in A_d. Obviously, this series converges for $s > 1$. The function $\zeta_K(s)$ represented in this way is called the Zeta-function of K. A careful selection of the system of representatives will enable us to compute this Zeta-function. For $a \in A_d$, $a \neq 0$, all elements associated to a are of the form $\pm a\epsilon^n$, $n \in \mathbb{Z}$. Every a is associated to an element $\pm a\epsilon^k$ which is located in the shaded area G in the figure below (the upper boundary is in G, the lower is not). One sees this by noticing that if one multiplies a by a unit $\pm \epsilon^k$, G maps to an area with $\pm \epsilon^k$ and $\pm \epsilon^{k+2}$ as boundaries.

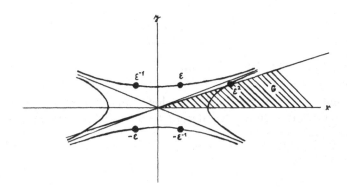

These areas are disjoint, $\mathbb{R}^2$ is their union. For $s > 1$ one obtains

$$\zeta_K(s) = \sum \frac{1}{\left(x^2 - dy^2\right)^s} \, ;$$

the summation extends over all $(x, y) \in G \cap A_d$, $(x, y) \neq (0,0)$. Similarly to the case treated on page 75, we can approximate $\zeta_K(s)$ by a double integral:

$$\sum \frac{1}{\left(x^2 - dy^2\right)^s} \approx 2 \int\int_H \frac{dx\,dy}{\left(x^2 - dy^2\right)^s} \, ,$$

where H denotes the area $\{(x, y) \in G \mid x^2 - dy^2 \geq 1\}$, i.e., the difference of the two expressions is bounded for $s \downarrow 1$. The factor 2 occurs because we consider intervals of area $\frac{1}{2}$ in the Riemann sum for the double integral since $d \equiv 1 \bmod 4$. We compute the integral by substituting $x \to \sqrt{d}\, z$. This transforms H into $H_1 = \{(z, y) \mid 0 \leq uy \leq v\sqrt{d}\, z,\ z^2 - y^2 \geq 1/d\}$, where $\epsilon^2 = u + v\sqrt{d}$, and we obtain

$$\iint_H \frac{dx\, dy}{(x^2 - dy^2)^s} = \frac{\sqrt{d}}{d^s} \iint_{H_1} \frac{dz\, dy}{(z^2 - y^2)^s}.$$

Let us now introduce the hyperbolic coordinates $z = r\cosh\theta$, $y = r\sinh\theta$. Then we obtain the following expression for our double integral

$$= \frac{\sqrt{d}}{d^s} \iint \frac{r\, dr\, d\theta}{r^{2s}};$$

where the integration is extended over $r \geq 1/\sqrt{d}$ and $0 \leq \theta \leq \log\epsilon^2$. This follows since the Jacobian

$$\begin{vmatrix} \dfrac{\partial z}{\partial r} & \dfrac{\partial z}{\partial \theta} \\[2mm] \dfrac{\partial y}{\partial r} & \dfrac{\partial y}{\partial \theta} \end{vmatrix}$$

equals

$$\begin{vmatrix} \cosh\theta & r\sinh\theta \\ \sinh\theta & r\cosh\theta \end{vmatrix} = r,$$

because $\cosh\theta = \frac{1}{2}(e^\theta + e^{-\theta})$, $\sinh\theta = \frac{1}{2}(e^\theta - e^{-\theta})$, $(\cosh\theta)^2 - (\sinh\theta)^2 = 1$. Our conditions $z^2 - y^2 \geq 1/d$ and $0 \leq uy \leq v\sqrt{d}\, z$ become $r \geq 1/\sqrt{d}$ and $ur\sinh\theta \leq vr\cosh\theta\sqrt{d}$ or $0 \leq \theta \leq \log\epsilon^2$ due to $\operatorname{arctanh}(x) = \frac{1}{2}\log(1 + x)/(1 - x)$. Let us continue to calculate the integral. We can now write it as

$$\frac{\sqrt{d}}{d^s} \log\epsilon^2 \int_{1/\sqrt{d}}^{\infty} r^{1-2s}\, dr$$

$$= \frac{\sqrt{d}}{d^s}\, 2\log\epsilon \left[\frac{r^{2-2s}}{2 - 2s} \right]_{1/\sqrt{d}}^{\infty}$$

$$= \frac{\sqrt{d}}{d^s}\, \frac{d^{s-1}}{2(s-1)}\, 2\log\epsilon = \frac{\log\epsilon}{\sqrt{d}}\, \frac{1}{(s-1)}.$$

Consequently,

$$\lim_{s \downarrow 1} (s - 1)\zeta_K(s) = \frac{2\log\epsilon}{\sqrt{d}}.$$

One has to modify the system of representatives of nonassociated elements in A_d when the fundamental unit has positive norm (we still assume $h = 1$).

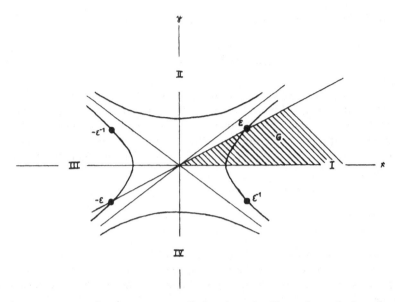

In that case, one chooses an area G that has the lines through 1 and ϵ as boundaries. Again, the upper boundary is part of G, and the lower is not. One obtains $\bigcup_{\eta\ \text{unit}} \eta G = \text{I} \cup \text{III}$, i.e., not the whole plane. This is the reason why one uses the transformation (having determinant 1)

$$\begin{bmatrix} 0 & -\sqrt{d} \\ 1/\sqrt{d} & 0 \end{bmatrix},$$

replacing x by $\sqrt{d}\,v$ and y by $-u/\sqrt{d}$. Obviously, this transformation does not change the integral

$$\int\int \frac{dx\,dy}{|x^2 - dy^2|^s} \; ;$$

so that in the case under consideration, one can approximate the Zeta-function of $\mathbb{Q}(\sqrt{d})$ by doubling (for $d \equiv 2, 3 \bmod 4$) or doubling twice (for $d \equiv 1 \bmod 4$) the value of the integral $\int\int_G [dx\,dy/(x^2 - dy^2)^s]$. For $K = \mathbb{Q}(\sqrt{d})$, $d > 0$, one obtains

$$\lim_{s\downarrow 1}(s-1)\zeta_K(s) = \begin{cases} \dfrac{2\log\epsilon}{\sqrt{d}} & \text{for } d \equiv 1 \quad \bmod 4, \\[2ex] \dfrac{\log\epsilon}{\sqrt{d}} & \text{for } d \equiv 2, 3 \quad \bmod 4 \end{cases}$$

$$= \frac{2\log\epsilon}{\sqrt{D}},$$

where D is the discriminant of K.

To establish the connection to the L-series, we write $\zeta_K(s)$ as an Euler product:

$$\zeta_K(s) = \prod_\pi \frac{1}{1 - |N(\pi)|^{-s}} :$$

π runs through a system of nonassociated prime elements of A_d. This representation of $\zeta_K(s)$ is possible since every element of A_d can be written uniquely (up to units) as the product of prime elements because of our assumption $h(K) = 1$. We now have to continue to evaluate the right-hand side; to this end, we will give a more precise description of the prime elements in A_d.

Step VII

Remark. Let $d \in \mathbb{Z}$ be square free and the class number of $\mathbb{Q}(\sqrt{d})$ be 1, i.e., in A_d unique prime factorization holds, and let $p \in \mathbb{Z}$ be a prime number. Then p can be decomposed into prime factors in A_d in the following way. Either

$$p = \epsilon\pi^2$$

with a unit ϵ and a prime element π in A_d (in this case, p is said to be ramified) or

$$p = \pm \pi\pi'$$

with nonassociated conjugate prime elements π, π' (in this case, p is said to be decomposed or split) or

$$p \text{ is a prime element}$$

in A_d (p is called inert in this case).

PROOF. Let π be a prime element in A_d of the form $\pi = x + y\sqrt{d}$ with $x, y \in \mathbb{Z}$ or $2x, 2y, x + y \in \mathbb{Z}$. Then $N(\pi) = \pi\pi' = x^2 - dy^2 = p_1 \ldots p_k$ with prime numbers $p_i \in \mathbb{Z}$. Without loss of generality we can assume that π is a divisor of $p = p_1$. Then $\pi\rho = p = \pi'\rho'$ with $\rho \in A_d$. Hence $\pi' | p$ and $\pi\pi' | p$ or $\pi\pi' | p^2$. In the first case, $\pi\pi' = \pm p$ and hence $p = \epsilon\pi^2$ if π, π' are associated or $p = \pm \pi\pi'$ if π, π' are nonassociated. In the second case, $\pi\pi' = \pm p^2 = \pm pp$. p is a prime element in A_d because one has unique prime factorization in A_d.

(8.9) **Theorem** (Euler, Gauss). *Under the same assumptions as in the preceding remark one has:*

(a) *p is ramified if and only if p divides the discriminant D of $\mathbb{Q}(\sqrt{d})$.*

(b) *If p is odd and relatively prime to D, then it splits if and only if $(\frac{d}{p}) = 1$; p is inert if and only if $(\frac{d}{p}) = -1$.*

(c) *If 2 does not divide D (i.e., $D \equiv 1 \bmod 4$), 2 splits if and only if $D \equiv 1 \bmod 8$ and is inert if and only if $D \equiv 5 \bmod 8$.*

PROOF. Assume $p \neq 2$ is a divisor of D or $p = 2$ and 2 is a divisor of d. Then p is a divisor of d. If $p = |d|$, then $p = \pm \sqrt{d} \sqrt{d}$, i.e., p is ramified in A_d. If $p < |d|$ one writes

$$d = p \frac{d}{p} = \sqrt{d} \sqrt{d} . \tag{*}$$

But p is not a divisor of $\sqrt{d}$ in A_d. Consequently, p is not a prime element in A_d. Then there is a prime element π in A_d with $\pi\pi' = \pm p$, π not a divisor of d/p. According to (*), π is also a divisor of $\sqrt{d}$, π^2 a divisor of d and hence a divisor of p, so p is ramified. If $p = 2$ is a divisor of D and not a divisor of d, then $d \equiv 3 \bmod 4$. One has

$$d^2 - d = 2 \frac{d^2 - d}{2} = (d + \sqrt{d})(d - \sqrt{d})$$

and $2 \nmid (d \pm \sqrt{d})$, hence 2 is not a prime element in A_d. This means that there is a prime element $x + y\sqrt{d}$ in A_d such that $\pm 2 = x^2 - dy^2$. Then it follows that

$$\pm \frac{x - y\sqrt{d}}{x + y\sqrt{d}} = \pm \frac{x^2 + dy^2 - 2xy\sqrt{d}}{x^2 - dy^2} = \frac{x^2 + dy^2}{2} - xy\sqrt{d}$$

and

$$\pm \frac{x + y\sqrt{d}}{x - y\sqrt{d}} = \frac{x^2 + dy^2}{2} + xy\sqrt{d}$$

are in A_d. Therefore $(x - y\sqrt{d})(x + y\sqrt{d})^{-1}$ is a unit, i.e., $x - y\sqrt{d}$ and $x + y\sqrt{d}$ are associated. Let $p \neq 2$ and relatively prime to D. If $(\frac{d}{p}) = 1$ then $x_0 \in \mathbb{Z}$ exists with

$$x_0^2 - d \equiv 0 \quad \bmod p.$$

Let us assume that p is a prime element of A_d. Then p is a divisor of $x_0 + \sqrt{d}$ or $x_0 - \sqrt{d}$. Then one of the two numbers $(x_0 + \sqrt{d})/p, (x_0 - \sqrt{d})/p$ is contained in A_d, which is a contradiction. This means that there is a prime element $x + y\sqrt{d}$ in A_d such that $\pm p = (x + y\sqrt{d})(x - y\sqrt{d}) = x^2 - dy^2$. Let us first assume that $d \not\equiv 1 \bmod 4$. Assume that $x + y\sqrt{d}, x - y\sqrt{d}$ are associated. Then $\pm (x + y\sqrt{d})/(x - y\sqrt{d}) = \pm (x^2 + dy^2 + 2xy\sqrt{d})/p$ is contained in A_d and p is consequently a divisor of x and y, which is a contradiction. We now consider the case $d \equiv 1 \bmod 4$. Assume that $x + y\sqrt{d}, x - y\sqrt{d}$ are associated. Then $\pm 4p = x_0^2 - dy_0^2$ with $x_0, y_0 \in \mathbb{Z}$ and, similar to the above, p is a divisor of x_0 and y_0, which is a contradiction. So we have shown that p is decomposed. Let $(\frac{d}{p}) = -1$. Assume that p is not a prime element in A_d. Then there is a prime element $x + y\sqrt{d} \in A_d$ with $\pm p = x^2 - dy^2$. Then $\pm 4p = (2x)^2 - d(2y)^2$, $(2x)^2 \equiv d(2y)^2 \bmod p$ with $2x, 2y \in \mathbb{Z}$. If p is a divisor of $2x$ or $2y$, then p^2 is a divisor of $4p$. So $2y$ and p are relatively prime. Then there is an integer z such that $2yz \equiv 1 \bmod p$, $(2xz)^2 \equiv d(2yz)^2 \equiv d \bmod p$, i.e., $(\frac{d}{p}) = 1$, which is a contradiction. If 2 is not a factor of D, then $d \equiv 1 \bmod 4$. If 2 is not a prime element in A_d, then there is a prime element $\pi = \frac{1}{2}(x + y\sqrt{d})$ in A_d such that

$$\pm 2 = \frac{1}{4}(x^2 - dy^2) \quad \text{or} \quad \pm 8 = x^2 - dy^2. \qquad (**)$$

If x and y are even, say $x = 2s$, $y = 2t$, then $s^2 - dt^2 = \pm 2$. But $s^2 - dt^2$ is either odd or a multiple of 4 because $d \equiv 1 \bmod 4$. Hence x and y are odd. Then $x^2 \equiv y^2 \equiv 1 \bmod 8$; from $(**)$ it follows that $x^2 - dy^2 \equiv 1 - d \equiv 0 \bmod 8$ and consequently $d \equiv 1 \bmod 8$. Hence 2 is a prime element in A_d if $d \equiv 5 \bmod 8$. Let $d \equiv 1 \bmod 8$. Then

$$\frac{1-d}{4} = 2\frac{1-d}{8} = \frac{1-\sqrt{d}}{2}\,\frac{1+\sqrt{d}}{2}$$

and $2 \nmid (1 \pm \sqrt{d})/2$. Since 2 is not a prime element in A_d, the equation $(**)$ is only satisfied for odd $x, y \in \mathbb{Z}$. The prime elements $\frac{1}{2}(x + y\sqrt{d}), \frac{1}{2}(x - y\sqrt{d})$ are not associated in A_d, for their quotient

$$\frac{x + y\sqrt{d}}{x - y\sqrt{d}} = \pm \frac{x^2 + dy^2}{8} \pm \frac{xy\sqrt{d}}{4}$$

is not in A_d.

The decomposition law can conveniently be expressed with the help of the character χ modulo D of $K = \mathbb{Q}(\sqrt{d})$ (which was defined on page 128) in the following way.

(8.10) Theorem. *The following conditions give the prime factorization of a rational prime number p in the ring of integers A_d of $\mathbb{Q}(\sqrt{d})$:*

(1) *p is ramified if and only if $\chi(p) = 0$.*
(2) *p is decomposed if and only if $\chi(p) = 1$.*
(3) *p is inert if and only if $\chi(p) = -1$.*

PROOF. Statement (1) is clear by the definition of a character modulo D. Let p not be a divisor of D. Then it suffices to show that $\chi(p) = (\frac{d}{p})$. We do this with the aid of the law of quadratic reciprocity. We need to distinguish between the cases $d \equiv 1$, 2 and 3 mod 4 and $d > 0$, $d < 0$. Take, for example, the case $d < 0$ and $d \equiv 3 \bmod 4$. By the law of quadratic reciprocity, one has

$$\left(\frac{d}{p}\right) = \left(\frac{-1}{p}\right)\prod_{q \mid d}\left(\frac{q}{p}\right) = (-1)^{(p-1)/2}\prod_{q \mid d}(-1)^{[(p-1)/2][(q-1)/2]}\left(\frac{p}{q}\right).$$

The number of prime numbers $\equiv 3 \bmod 4$ which are factors of d is even. The last expression can be written as

$$= (-1)^{(p-1)/2}\prod_{q \mid d}\left(\frac{p}{q}\right) = \chi(p).$$

The other cases are treated similarly. We illustrate this with another example, $d > 0$, $d = 2\delta \equiv 2 \bmod 4$. Again, by the law of quadratic reciprocity, one has

$$\left(\frac{d}{p}\right) = \left(\frac{2\delta}{p}\right) = \left(\frac{2}{p}\right)\prod_{q\mid\delta}\left(\frac{q}{p}\right) = (-1)^{(p^2-1)/8}\prod_{q\mid\delta}(-1)^{[(p-1)/2][(q-1)/2]}\left(\frac{p}{q}\right).$$

The number of prime numbers $\equiv 3 \bmod 4$ which are factors of δ is $\equiv [(\delta - 1)/2] \bmod 2$. This means that the last expression can be written as

$$= (-1)^{(p^2-1)/8}(-1)^{[(p-1)/2][(\delta-1)/2]}\prod_{q\mid\delta}\left(\frac{p}{q}\right) = \chi(p).$$

Step VIII

For the Zeta-function, the decomposition law means the following:

$$\zeta_K(s) = \sum_{(a)\neq 0}|N(a)|^{-s} = \prod_{\pi}\frac{1}{1 - |N(\pi)|^{-s}}$$

$$= \prod_{\pi\mid p}\frac{1}{1 - p^{-s}}\prod_{\pi\mid p}\left(\frac{1}{1 - p^{-s}}\right)^2\prod_{\pi\mid p}\frac{1}{1 - p^{-2s}}$$

$$\quad\; p \text{ ramified}\qquad p \text{ decomposed}\qquad p \text{ inert}$$

$$= \prod_{\chi(p)=0}\frac{1}{1 - p^{-s}}\prod_{\chi(p)=1}\left(\frac{1}{1 - p^{-s}}\right)^2\prod_{\chi(p)=-1}\left(\frac{1}{1 - p^{-2s}}\right)$$

$$= \zeta(s)\prod_{\chi(p)=1}\frac{1}{1 - p^{-s}}\prod_{\chi(p)=-1}\frac{1}{1 + p^{-s}}$$

$$= \zeta(s)L(s,\chi),$$

where $L(s,\chi) = \sum_{n=1}^{\infty}\chi(n)n^{-s}$ is the L-function belonging to the character χ of $K = \mathbb{Q}(\sqrt{d})$. Since

$$\lim_{s\downarrow 1}(s - 1)\zeta_K(s) = \frac{2\log\epsilon}{\sqrt{D}}\qquad (\text{for } d > 0),$$

one has

$$L(1,\chi) = \frac{2\log\epsilon}{\sqrt{D}}.$$

In general, one can show:

(8.11) **Theorem.** *For $d > 0$,*

$$L(1,\chi) = \frac{\log\rho}{\sqrt{D}}\qquad with\quad \rho = \frac{\prod\sin(\pi b/D)}{\prod\sin(\pi a/D)},$$

where b runs through the set $\{k \mid 1 \leqslant k \leqslant D, \chi(k) = -1\}$ *and a through the set* $\{k \mid 1 \leqslant k \leqslant D, \chi(k) = 1\}$.

For $D = d = p \equiv 1 \mod 4$, we have already proved this theorem. The general case does not pose any new essential difficulties. This theorem has a very interesting consequence.

(8.12) Corollary. *Let* $d > 0$ *and suppose the class number of* $\mathbb{Q}(\sqrt{d})$ *is* 1. *Then*

$$\epsilon^2 = \frac{\prod \sin(\pi b / D)}{\prod \sin(\pi a / D)} \qquad (b, a \text{ as in } (8.11))$$

for the fundamental unit ϵ *of* A_d.

This is an explicit formula for the fundamental unit ϵ in which all factors occur twice because of the periodicity of the sine function. We now have two methods for the determination of the fundamental unit, continued fractions and a representation by a trigonometric function. It is hard to believe that both lead to the same answer. Let us look at an example, $d = 13$. The fundamental unit is $(3 + \sqrt{13})/2 = 3.302775636 \ldots$. The quadratic non-residues in the interval $[1, 13]$ are $2, 5, 6, 7, 8, 11$, and the quadratic residues $1, 4, 9, 3, 12, 10$. Hence

$$\epsilon = \frac{\sin(\pi 2 / 13)\sin(\pi 5 / 13)\sin(\pi 6 / 13)}{\sin(\pi / 13)\sin(\pi 3 / 13)\sin(\pi 4 / 13)},$$

which is indeed

$$3.302775636 \ldots .$$

To this point we have not considered the case $d < 0$ (except for $d = -1, -3$ in Chapter 6). We mention the relevant results now, still excluding the cases $d = -1, -3$. For $d \equiv 2, 3 \mod 4$ one has for the Zeta-function $\zeta_K(s)$ of $K = \mathbb{Q}(\sqrt{d})$, $d < 0$ the following expressions if $h(k) = 1$.

$$\zeta_K(s) = \frac{1}{2} \sum_{(x,y) \neq (0,0)} \frac{1}{(x^2 - dy^2)^s} \approx \frac{1}{2} \int \int_{x^2 - dy^2 \geq 1} \frac{dx\, dy}{(x^2 - dy^2)^s}$$

$$= \frac{1}{2} \int \int_{-dz^2 - dy^2 \geq 1} \frac{d(\sqrt{-d}\, z)\, dy}{(-dz^2 - dy^2)^s} = \frac{\sqrt{|d|}}{2|d|^s} \int \int_{z^2 + y^2 \geq 1/|d|} \frac{dz\, dy}{(z^2 + y^2)^s}$$

$$= \frac{1}{2} \frac{1}{\sqrt{|d|^{2s-1}}} \int \int_{r^2 \geq 1/|d|} r^{-2s} r\, dr\, d\theta = \frac{2\pi}{\sqrt{|d|^{2s-1}}} \left[\frac{r^{2-2s}}{2 - 2s} \right]_{1/\sqrt{|d|}}^{\infty}$$

$$= \frac{1}{2} \frac{\pi}{s - 1} \frac{|d|^{s-1}}{\sqrt{|d|^{2s-1}}}$$

and hence

$$\lim_{s \downarrow 1} (s - 1) \zeta_K(s) = \frac{\pi}{2\sqrt{|d|}} \, .$$

For the case $d \equiv 1 \bmod 4$, one obtains

$$\lim_{s \downarrow 1} (s - 1) \zeta_K(s) = \frac{\pi}{\sqrt{|d|}} \, ,$$

i.e., always

$$\lim_{s \downarrow 1} (s - 1) \zeta_K(s) = \frac{\pi}{\sqrt{|D|}} \, .$$

This leads to

$$L(1, \chi) = \frac{\pi}{\sqrt{|D|}}$$

with

$$L(1, \chi) = \frac{\pi}{|D|\sqrt{|D|}} \left| \sum_{k=1}^{|D|-1} \chi(k)k \right| .$$

(See also page 126.)

Step IX

Up until now, we have always assumed that the class number of the field $K = \mathbb{Q}(\sqrt{d})$ equals 1. Nevertheless, we have essentially completed the analytical part of the proof of the class number formula on page 128. We did not, however, cover several important algebraic facts which we now mention without proof. One of the fundamental steps was to represent the Zeta-function of $\mathbb{Q}(\sqrt{d})$ as a Euler product. This was based on the unique decomposition of the elements of A_d as a product of prime elements. This decomposition is not possible in general. Kummer found a way around it (see the historical remarks in Edwards, *Fermat's Last Theorem*, page 76) by introducing his so-called ideal numbers or, as one says today, ideals. An ideal I in a commutative ring R with unit is a subgroup of the additive group of R such that $ra \in I$ for all $r \in R$, $a \in I$. I is a prime ideal if $I \neq R$ and $ab \in I$ implies either a or $b \in I$, i.e., the quotient ring R/I has no zero divisors. For two ideals I, J in R, one defines the product IJ as the ideal generated by $\{ab \mid a \in I, b \in J\}$, i.e.,

$$IJ := \left\{ \sum a_i b_i c_i \mid a_i \in I, b_i \in J, c_i \in R \right\}.$$

The following basic theorem holds.

(8.13) **Theorem.** *In the ring A_d, every ideal $\mathfrak{a} \neq 0$ can be written uniquely as* *a product of prime ideals* $\mathfrak{p}_i$,

$$\mathfrak{a} = \mathfrak{p}_1^{e_1} \ldots \mathfrak{p}_k^{e_k}.$$

The norm $N(\mathfrak{a})$ of an ideal $\mathfrak{a}$ in A_d is defined as the number of elements in the residue class ring $A_d/\mathfrak{a}$,

$$N(\mathfrak{a}) := |A_d/\mathfrak{a}|.$$

$A_d/\mathfrak{a}$ is finite. Also, $N(\mathfrak{ab}) = N(\mathfrak{a})N(\mathfrak{b})$. The fact that every ideal $\mathfrak{a}$ in A_d, $\mathfrak{a} \neq 0$, can be decomposed uniquely into a product of prime ideals and the product formula for the norm allow us to write the Zeta-function of $K = \mathbb{Q}(\sqrt{d})$ as an Euler product for $s > 1$,

$$\zeta_K(s) = \sum_{\mathfrak{a} \neq 0} N(\mathfrak{a})^{-s} = \prod_{\mathfrak{p}} \frac{1}{1 - N(\mathfrak{p})^{-s}}.$$

The decomposition law can directly be transferred into the form of (8.10): if p is a rational prime number, generating the principal ideal $(p) = pA_d$ and if χ is the character of A_d, then

$$\begin{aligned}
(p) &= \mathfrak{p}^2 & \text{if and only if} \quad \chi(p) &= 0 & (p \text{ ramified}),\\
(p) &= \mathfrak{p}\mathfrak{p}' & \text{if and only if} \quad \chi(p) &= 1 & (p \text{ splits}),\\
(p) &= \mathfrak{p} & \text{if and only if} \quad \chi(p) &= -1 & (p \text{ inert}).
\end{aligned}$$

Similar to the way we proceeded on page 139, one shows that, for $s > 1$,

$$\zeta_K(s) = \zeta(s)L(s,\chi)$$

for some L-function $L(s,\chi)$. Hence

$$\lim_{s \downarrow 1} (s-1)\zeta_K(s) = L(1,\chi).$$

One gets a connection with the class number h of $\mathbb{Q}(\sqrt{d})$, that is the order of the class group C, by writing

$$\zeta_K(s) = \sum_{c \in C} \zeta_c(s)$$

with

$$\zeta_c(s) = \sum_{\mathfrak{a} \in c} N(\mathfrak{a})^{-s}.$$

Then one can prove:

(8.14) Lemma. $\lim_{s \downarrow 1}(s-1)\zeta_c(s)$ *is independent of* c.

PROOF. Let $\mathfrak{a}' \in c^{-1}$. Then $\mathfrak{a}' = (\alpha)$ is an integral ideal and

$$\zeta_c(s) = \sum_{\mathfrak{a} \in c} N(\mathfrak{a})^{-s} = N(\mathfrak{a}')^s \sum_{\substack{\alpha \neq 0 \\ \alpha \in \mathfrak{a}' \\ \text{not assoc.}}} N(\alpha)^{-s}.$$

A_d is the disjoint union of $N(\mathfrak{a}')$ residue classes $\beta + \mathfrak{a}'$. It is easy to show that

$$\sum_{\substack{\alpha \neq 0 \\ \alpha \in \beta + \mathfrak{a}' \\ \text{not assoc.}}} N(\alpha)^{-s} \approx \sum_{\substack{\alpha \neq 0 \\ \alpha \in \mathfrak{a}' \\ \text{not assoc.}}} N(\alpha)^{-s}.$$

Hence

$$\zeta_c(s) = N(\mathfrak{a}')^s \sum_{\substack{\alpha \neq 0 \\ \alpha \in \mathfrak{a}' \\ \text{not assoc.}}} N(\alpha)^{-s} \approx \frac{N(\mathfrak{a}')^s}{N(\mathfrak{a}')} \zeta_1(s),$$

which proves our assertion.

Choosing for c the principal class A_d and using earlier results (see pages 139 and 141),

$$L(1, \chi) = \lim_{s \downarrow 1} (s - 1)\zeta_K(s) = \begin{cases} h \dfrac{2 \log \epsilon}{\sqrt{D}} & \text{for } d > 0, \\[3mm] h \dfrac{\pi}{\sqrt{|D|}} & \text{for } d < 0 \end{cases}$$

with our usual L-function $L(s, \chi)$.

This completes our sketch of a proof of the class number formula. For details, we refer the reader to the literature. We think it has become clear that a whole series of number-theoretical questions is connected to the problem of the class number formula, a fact that is not obvious from the formula itself. We have also seen how deeply one can penetrate with the help of various analytical methods. Moreover, with the aid of these methods, one obtains a number of surprising and interesting results. It is obvious that Dirichlet's methods opened up a completely new perspective on number theory. His results will always be among the most important work ever done in mathematics.

In the first third of the nineteenth century, the bourgeoisie became the main exponent of cultural life. This development was initiated by the French revolution: the emerging industrialization and the social changes that took place paved its way. Neohumanism with its emphasis on the ideals of classical antiquity formed the basis of intellectual life in Germany.

Wilhelm and Alexander von Humboldt's reformation of the Prussian universities had a lasting influence on scientific organization in Germany. Trying to express the spirit and the ideals of this time, we can do no better than to quote E. E. Kummer in his memorial to Dirichlet (Gedächtnisrede auf Gustav Peter Lejeune Dirichlet):

Gustav Peter Lejeune Dirichlet

Gustav Peter Lejeune Dirichlet was born in Düren on February 13, 1805. His father was a postmaster, a gentle, pleasant, and civil man; his mother, a spirited and well-educated woman, gave the exceptionally gifted boy a very careful education. ... His parents wished that he would be a merchant but since he showed a decided aversion to this profession they changed their minds and sent him to Bonn to the Gymnasium in 1817. ... He was conspicuous by his decency and good manners; the ease and openness of his character made everybody who came into contact with him his friend. He was industrious but his main efforts were dedicated to mathematics and history. He studied even if he did not have any homework from school because his active mind was always occupied with worthy objects of thought. Great historical events, like the French revolution, and public affairs interested him deeply. He judged these and other things with an independence unusual for his youth, and from a liberal viewpoint, probably the fruit of his parents' teaching. Dirichlet stayed at the Gymnasium in Bonn for only two years and subsequently moved to the Jesuit gymnasium in Cologne. His teacher in mathematics was Georg Simon Ohm, later famous for the law of electrical resistance. Ohm's instruction, as well as his assiduous independent study of mathematical works, helped him to make remarkable progress in this science and to acquire an unusual degree of knowledge. He completed the course at the Gymnasium very quickly and obtained in 1821, when he was only 16 years old, the final diploma which permitted him to go to University. When he went home to discuss the choice of his future profession with his parents it was natural that they seriously questioned his decision to study mathematics and admonished him to secure his place in the world by choosing a more practically oriented subject, the law. Modestly but firmly he declared that if they requested this from him he would agree but he was not able to give up his favorite subject and would at least devote his nights to it. His parents, who were as rational as they were tenderly loving, gave in to their son's request.

At that time, mathematical education at the Prussian and the other German universities was at a low point. The lectures, whose level barely exceeded that of elementary mathematics, were in no way capable of satisfying the desire for knowledge which burned in the young Dirichlet; also, except for the great Gauss, there was nobody in Germany whose name could have attracted him. In France, and particularly in Paris, mathematics was still in full bloom; a circle of men, whose names will shine in the history of mathematics for all time, worked as researchers and teachers and contributed to the development and propagation of our science. ... Judging these circumstances correctly, Dirichlet decided that Paris would be the place where he could expect the greatest gain for his mathematical studies and went to this center of mathematical sciences in May of 1822, in the happy expectation that he could now wholly devote himself to his favorite subject. He listened to the lectures at the Collège de France and the Faculté des Sciences. ... Besides attending these lectures and working through the material, Dirichlet devoted his time to the attentive study of the most eminent mathematical works and among those specifically Gauss's works on higher arithmetic, *Disquisitiones Arithmeticae*. This book exercised a more important and deeper influence on his mathematical education and the direction of his interests than all his other work in Paris. He went through it not only more than once but he never ceased, throughout his life, to return again and again to its wealth of deep mathematical ideas. This is the reason why it was never on his bookshelf but had a permanent place on his desk. ...

Dirichlet's life during the first years of his stay in Paris was most simple and withdrawn. ... This changed during the Summer of 1823 which was of greatest importance to his general education. General Foy was a universally educated man, a leader excelling through his prominent station as head of the opposition in the House of Deputies, as well as being one of its most celebrated speakers and his illustrious military career. His house was one of the most renowned and sought out in Paris; at that time, the General looked for a young man to be a teacher of his children ... and Dirichlet was recommended to him through the good offices of a friend. During the first personal interview the open and modest character of the young man made such a good impression on the General that he immediately offered him the position, with a decent salary and with such small obligations that there was enough free time for Dirichlet to continue his studies. But the General exercised his deepest influence by his example of an active, noble, and well-educated man. This influence extended not only to Dirichlet's education, his manners, and preferences, but also to his way of thinking and acting and his general outlook on life. It was also of great importance that the house of the General, which was the center of the most prominent artists and scientists in the capital of France, gave Dirichlet the opportunity to look at life on a grand scale and to actively participate in it. All these new impressions ... did not distract Dirichlet from his mathematical studies; to the contrary, during this time he assiduously worked on the first of his published papers, Memoire sur l'impossibilité de quelques équations indéterminées du cinquième degré. ... Not only through its new results, gained in one of the most difficult parts of number theory, but also by its conciseness and sharpness of the proofs and the exceptional clarity of presentation this first paper secured for Dirichlet an illustrious success. . . . It established Dirichlet's reputation as an excellent mathematician; as a young man with a great future he was not only admitted to the highest scientific circles in Paris, he was sought out by them. He came into close connection with several of the most reputable members of the Academy in Paris among whom two have to be singled out, Fourier, who influenced the direction of Dirichlet's future scientific investigations, and Alexander von Humboldt, who influenced the further development of his life.

As a young man, Fourier participated at the foundation of the Ecole Normale and Ecole Polytechnique; he retained his enthusiasm for active scientific communication, and he had an inner need to tell of the beautiful and great matters which he investigated. In Dirichlet, he found a young man . . . by whom he was not only admired but also completely understood.

Alexander von Humboldt, who at that time lived in Paris . . . , besides respecting his talents and his scientific skills, also gave him his most vivid personal attention and affection which he continued to harbor and to express. Already, during this first visit to Paris, Dirichlet expressed, in a course of a conversation, his intention to return later to his native country. Humboldt . . . confirmed him in this plan. The death of his sponsor, General Foy, in November 1825, and the influence of Alexander von Humboldt, who soon after left Paris and moved to Berlin, prompted Dirichlet to realize his intention of returning to his native country. He petitioned the Minister von Altenstein for suitable employment which Humboldt undertook to second and to further by his influence. He was very active in the matter of Dirichlet's position, but all his efforts, which Gauss himself assisted by a letter to our colleague, Mr. Encke, who forwarded it to the Royal Ministry, could not achieve more than the assurance of 400 Taler as remuneration to help Dirichlet to become a Privatdozent in Breslau. Since this remuneration assured him a modest living and since he could rely on Humboldt's efforts to help him find a more suitable position, Dirichlet accepted without further consideration. In the meantime, he had been made Doctor of Philosophy *honoris causa* by the philosophical faculty of the University of Bonn, which made it considerably easier to become a Privatdozent at a university.

On his way to Breslau, Dirichlet chose to go via Göttingen to meet Gauss personally. He visited him on March 18, 1827. I did not find any details about this meeting, but in a letter to his mother, Dirichlet says that Gauss received him very kindly and that his personal impression of this great man was much better than expected.

In the meantime, Alexander von Humboldt had succeeded in having Dirichlet named ausserordentlicher Professor at the University of Breslau. He now went to work to gain him for the University and Academy here, but first to get him to Berlin. A suitable occasion was a vacancy at the Allgemeine Kriegsschule; Humboldt seized this opportunity and recommended Dirichlet very highly to the General von Radowitz and the Minister of War. There was, however, no immediate firm offer, probably because Dirichlet was only 23 years old and might have seemed to be too young for such a position; consequently, the Minister of Altenstein was asked to agree that Dirichlet would be granted a sabbatical of one year to be a substitute teacher at the Kriegsschule.

In the fall of 1828 he came to Berlin to start at his new position. He was very fond of giving these lectures which he had to present to officers who were approximately contemporaries of his . . . .

Soon after his arrival in Berlin, Dirichlet undertook the necessary steps to be permitted to give lectures at the University. As Professor at another university, he was not entitled to do this and there was no alternative to being a Privatdozent once more. He consequently directed such a petition to the Philosophical Faculty. Since he already had proved his scientific competence he was immediately permitted to give lectures with the title of a Privatdozent. Only in 1831 was he formally made ausserordentlicher Professor at the University here; a few months later he was made a full member of our Academy. In the same year, he married Rebecca Mendelsohn-Bartholdy, a granddaughter of Moses Mendelsohn. It is remarkable that Alexander von Humboldt had a share even in this, because he introduced Dirichlet to the house of his future in-laws, which was so famous for its intellectual spirit and love of the arts.

After this his life is of less interest than his scientific works which Dirichlet prepared during the next 27 years.

Reading these words of Kummer today and comparing them to our own attitude—much more skeptical and in its values less certain—the difference is obvious. But it is edifying to follow the undramatic but remarkable career of this young man who was equipped with firm ideals and convictions, overpowering mental facilities, and a winning and kind character—this is emphasized again and again. In just two years he finds access, despite his petit-bourgeois origins, to the highest circles of bourgeois society.

The life of Nils Hendrick Abel, intellectually Dirichlet's peer and a person as appealing as Dirichlet, is a reminder of just how fortunate Dirichlet was. Dirichlet and Abel had met in Paris and held each other in high regard, but they did not stay in contact.

At this point we would like to add a few words about Alexander von Humboldt, whose positive political influence can hardly be overestimated. The German University which only now, more than 150 years later, has entered a new phase, was basically established by Humboldt. With great personal effort, Humboldt helped the careers of many scientific talents, among them most of the mathematicians of his time, like Gotthold Eisenstein, who, sickly and depressed, had barely any personal contacts and whom, despite an age difference of nearly 50 years, von Humboldt aided again and again.

In his memorial, Kummer discusses Dirichlet's scientific work extensively. There is no need to repeat the passages on number theory which have been described in this chapter. A few sentences will conclude Dirichlet's biography. Until 1855 he stayed in Berlin, at the center of a growing circle of important colleagues and students, Jacobi, Steiner, Borchardt, Kummer, Eisenstein, Kronecker, Dedekind, and Riemann, to name the most important among them. A close scientific and personal friendship of more than a quarter of a century tied him to Jacobi. Kummer writes:

The common interest in the knowledge of truth and the furtherance of the mathematical sciences was the firm foundation of the friendly relations between Jacobi and Dirichlet. They saw each other virtually every day and

Carl Gustav Jacob Jacobi

discussed more general or specific scientific questions; the difference between their perspectives of the whole of mathematics lent their spirited discussions a vivid and always renewed interest. Jacobi, who on account of the wonderful broadness of mind no less than by the depth of his mathematical investigations and the glamor of his discoveries gained acceptance and acknowledgment everywhere, enjoyed a wider reputation than Dirichlet, who did not know how to build a reputation; his papers, which treated only the most difficult problems of our science, had a smaller circle of readers and admirers. Nobody was more aware of the odd disparity between the reputation and scientific importance of Dirichlet than Jacobi; nobody was more skillful and more active at counterbalancing this and to seek for his friend the most deserved acknowledgement in wider circles.

Dirichlet's wife gives a more personal account of the relations with the so different, extroverted, lively, ironical, and aggressive Jacobi, after whose death she writes:

> . . . enough that he died and that the world has lost a gigantic mind who was so close to us with all his mistakes and virtues. His relation to Dirichlet was nice—for hours they sat together, calling it "to be silent about mathematics" ("Mathematikschweigen"); they never spared each other and Dirichlet often told him the bitterest truths, but Jacobi understood this well and he made his great mind bend before Dirichlet's great character . . .

After Gauss's death in 1855, the university in Göttingen "which for half a century had been proud to own the first of all living mathematicians,

Ernst Eduard Kummer

tried to preserve this glory by inviting Dirichlet to occupy Gauss's chair"
(Kummer). He accepted the offer, since the more quiet atmosphere in Göttingen was very much to his taste. In Berlin, the considerable teaching obligations at the Kriegsschule were often a great aggravation; in Göttingen,
he had more time. But only four years remained for him; he died after a

heart attack on May 5, 1853. Dirichlet has not published much; neither did he leave much unpublished material even though it is known that he was mathematically active into his last year. He was used to keeping his thoughts to himself and was reluctant to write down his results. However, the relatively little that has been published is among the most perfect and important contributions to mathematics.

References

G. P. L. Dirichlet: *Werke*, particularly the papers: Beweis des Satzes, dass jede unbegrenzte arithmetische Progression, deren erstes Glied und Differenz ganze Zahlen ohne gemeinschaftlichen Faktor sind, unendlich viele Primzahlen enthält, 1837. Sur la manière de résoudre l'équation $t^2 - pn^2 = 1$ au moyen des fonctions circulaires, 1837. Recherches sur diverses applications de l'analyse infinitésimale à la théorie des nombres, 1839/40.

O. Ore: Dirichlet, Gustav Peter Lejeune in *Dictionary of Scientific Biography*.

E. Kummer: *Gedächtnisrede auf Gustav Peter Lejeune Dirichlet, Dirichlet's Works* or *Kummer's Works*.

H. Minkowski: Peter Gustav Lejeune Dirichlet und seine Bedeutung für die heutige Mathematik. *Gesammelte Abhandlungen* II.

K.-R. Biermann: *Johann Peter Gustav Lejeune Dirichlet, Dokumente für sein Leben und Wirken*. Abh. Deutsche Akad. Wiss. Berlin, Klasse für Mathematik, Physik und Technik, Akademie-Verlag, Berlin, 1959.

CHAPTER 9

From Hermite to Minkowski

In Chapter 6 we saw that the theory of binary quadratic forms is essentially equivalent to the theory of quadratic number fields. After Gauss, number theory developed in two basically different directions, the theory of algebraic number fields, i.e., finite extensions of $\mathbb{Q}$ as generalizations of quadratic number fields, and the theory of (integral) quadratic forms in several variables and their automorphisms, as a generalization of binary quadratic forms. In this chapter, we will sketch the development of certain aspects of the latter. To do this, we have to introduce a few basic concepts; for the sake of simplicity, we will use modern terminology.

A (*symmetric*) *bilinear space* over $\mathbb{Z}$ consists of a pair (N, b), where N is a finitely generated free $\mathbb{Z}$-module and $b : N \times N \to \mathbb{R}$ a $\mathbb{Z}$-bilinear symmetric mapping. We define the quadratic form $q(x) = b(x, x)$. This completely determines the bilinear form b because $b(x, y) = \frac{1}{2}(q(x + y) - q(x) - q(y))$.

Let $e_1, \ldots, e_n$ be a basis of N and $x = \sum_{i=1}^{n} x_i e_i$, $y = \sum_{j=1}^{n} y_j e_j$. Then

$$b(x, y) = \sum_{i,j} x_i b(e_i, e_j) y_j$$

$$= (x_1, \ldots, x_n)(b(e_i, e_j))(y_1, \ldots, y_n)^t$$

$$= xBy^t,$$

where B denotes the symmetric matrix $(b(e_i, e_j))_{i,j} \in M(n, \mathbb{R})$ and t the transpose. Conversely, every symmetric matrix $B \in M(n, \mathbb{R})$ defines a bilinear space $(\mathbb{Z}^n, b)$ by

$$b(x, y) := xBy^t$$

for all $x = (x_1, \ldots, x_n)$, $y = (y_1, \ldots, y_n) \in \mathbb{Z}^n$. Two bilinear spaces (N, b),

(N', b') are isomorphic if there is a $\mathbb{Z}$-linear isomorphism $\alpha : N \to N'$ with

$$b'(\alpha x, \alpha y) = b(x, y) \qquad \text{for all} \quad x, y \in N.$$

One of the basic problems is to determine all isomorphism classes of bilinear spaces.

Let us interpret isomorphisms using matrices. Let B, B' be matrices belonging to $(N, b), (N', b')$ and $A \in GL(n, \mathbb{Z})$, a matrix belonging to α. Then the isomorphism $(N, b) \cong (N', b')$ means

$$xBy^t = (xA)B'(yA)^t = xAB'A^t y^t,$$

that is

$$B = AB'A^t.$$

In general, we call symmetric matrices B, B' congruent if they satisfy this relation. Expressed in the language of matrices, our main problem is to determine all congruence classes of symmetric matrices.

The *determinant* of a bilinear space (N, b) is the determinant of a matrix belonging to b. The determinant of (N, b) is indeed uniquely defined because a change of basis, with $A \in GL(n, \mathbb{Z})$, transforms the matrix B to the matrix ABA^t; then $\det(ABA^t) = \det(B)\det(A)^2 = \det(B)$. If one allows rings other than $\mathbb{Z}$, then the determinant of a bilinear space is determined only up to squares of units.

Two elements $x, y \in N$ are called *orthogonal* (written $x \perp y$) if $b(x, y) = 0$. The submodule

$$X^\perp := \{ y \in N \mid y \perp x \text{ for all } x \in X \}$$

is called the *orthogonal complement* of a submodule $X \subset N$ (with respect to b). Suppose that $(N_1, b_1), \ldots, (N_t, b_t)$ are bilinear spaces over $\mathbb{Z}$; then $\perp_{i=1}^{t}(N_i, b_i)$ is the bilinear space (N, b) with $N = \bigoplus_{i=1}^{t} N_i$ and

$$b(x_1 \oplus \cdots \oplus x_t, y_1 \oplus \cdots \oplus y_t) = \sum_{i=1}^{t} b_i(x_i, y_i).$$

$\perp_{i=1}^{t}(N_i, b_i)$ is called the *orthogonal sum* of (N_i, b_i). Using matrices, $(N, b) = \perp_{i=1}^{t}(N_i, b_i)$ means that a matrix belonging to (N, b) is congruent to a matrix of the form

$$\begin{bmatrix} B_1 & & & 0 \\ & B_2 & & \\ & & \ddots & \\ 0 & & & B_t \end{bmatrix},$$

where the B_i are matrices for (N_i, b_i). A symmetric bilinear form $b : N \times N \to \mathbb{R}$ is called *positive definite* if $b(x, x) > 0$ for all $x \neq 0$. One makes an analogous definition for symmetric matrices. Generalizing the reduction theorem (4.2) of Lagrange in the binary case, Charles Hermite (1822–1901) proved the following theorem.

(9.1) Theorem. *Let (N,b) be a positive bilinear space of rank n with determinant D. Then there exists $x \in N$ with*

$$0 < b(x,x) \leqslant \left(\frac{4}{3}\right)^{(n-1)/2}\sqrt[n]{D} .$$

PROOF (By induction on n). Without loss of generality we can assume $N = \mathbb{Z}^n$. The case $n = 1$ is clear. For $n > 1$, choose $e_1 \in \mathbb{Z}^n$ such that $M = b(e_1, e_1)$ is minimal. Extend b bilinearly to all of $\mathbb{R}^n$ and consider the orthogonal projection π on the hyperplane

$$H = \{\, y \in \mathbb{R}^n \mid b(e_1, y) = 0 \},$$

i.e., the mapping $\pi : \mathbb{R}^n \to H$ with $\pi(e_1) = 0$, $\pi|_H = $ identity; clearly $\pi(x) = x - [b(x, e_1)/b(e_1, e_1)]e_1$. Let $e_1, \ldots, e_n$ be a basis of $\mathbb{Z}^n$ and $L := \mathbb{Z}\pi(e_2) + \cdots + \mathbb{Z}\pi(e_n)$. L has dimension $n - 1$. Use the matrix

$$A = \begin{bmatrix} 1 & \alpha_2 & \cdots & \alpha_n \\ & & \ddots & 0 \\ & 0 & & \ddots \\ & & & 1 \end{bmatrix}$$

with $\alpha_i = -b(e_i, e_1)/b(e_1, e_1)$ to switch from the basis $e_1, \ldots, e_n$ to the basis $e_1, \pi(e_2), \ldots, \pi(e_n)$. With $B = (b(e_i, e_j))$, one obtains

$$ABA^t = \left(\begin{array}{c|c} b(e_1, e_1) & 0 \\ \hline 0 & B' \end{array}\right),$$

where B' is the matrix belonging to $b|_{L \times L}$. Then $D = Md$ with $d = \det((L, b|_{L \times L}))$. By induction there is $x \in L$, $x \neq 0$, such that

$$b(x,x) \leqslant \left(\frac{4}{3}\right)^{(n-2)/2}\sqrt[n-1]{d} .$$

Let $y \in \mathbb{Z}^n$ be such that $\pi(y) = x$. Then $\pi(y) = y + te_1$ with $t \in \mathbb{R}$. By adding, if necessary, a suitable integral multiple of e_1 to y, we can ensure that $|t| \leqslant \frac{1}{2}$. This is the decisive step of the proof; for then

$$M = b(e_1, e_1) \leqq b(y, y) = b(x - te_1, x - te_1)$$

$$= b(x,x) + t^2 b(e_1, e_1) \leqq b(x,x) + \tfrac{1}{4}M,$$

and hence

$$\tfrac{3}{4}M \leqslant b(x,x) \leqq \left(\frac{4}{3}\right)^{(n-2)/2}\sqrt[n-1]{d} ,$$

$$M \leqq \left(\frac{4}{3}\right)^{n/2}\sqrt[n-1]{\frac{D}{M}} ,$$

$$M^{n/(n-1)} \leqq \left(\frac{4}{3}\right)^{n/2}\sqrt[n-1]{D} ,$$

$$M \leqq \left(\frac{4}{3}\right)^{(n-1)/2}\sqrt[n]{D} .$$

Hermite communicated his result to Jacobi in a letter of August 6, 1845. He stated "de nombreuses questions me semblent dépendre des resultats précédents." (Numerous questions seem to depend on the preceding results.) Concluding this letter, he derived several results of Jacobi with the help of his estimate. In a second letter to Jacobi, he discussed further applications; we will come to them presently.

For now, we consider only bilinear spaces over $\mathbb{Z}$ which assume integral values, i.e., bilinear forms $b : N \times N \to \mathbb{Z}$. Such bilinear spaces are called *unimodular* or *nonsingular* if the determinant equals ± 1.

Lemma. *Let (N,b) be a bilinear space over $\mathbb{Z}$, M a nonsingular submodule of N. Then (N,b) is isomorphic to $(M,b) \perp (M^{\perp}, b)$.*

PROOF. Let $e_1, \ldots, e_k$ be a basis of M which can be extended to a basis $e_1, \ldots, e_n$ of N. Then $B = (b(e_i, e_j))$, $1 \leqslant i, j \leqslant n$ has the form

$$B = \left(\begin{array}{c|c} B_1 & C \\ \hline C' & D \end{array} \right)$$

with $B_1 = (b(e_i, e_j))$, $1 \leqslant i, j \leqslant k$. By our assumption, B_1 is invertible in the ring $M(n, \mathbb{Z})$ of matrices with integral entries. One has

$$\begin{pmatrix} E & 0 \\ -(B_1^{-1}C)' & E \end{pmatrix} \begin{pmatrix} B_1 & C \\ C' & D \end{pmatrix} \begin{pmatrix} E & -B_1^{-1}C \\ 0 & E \end{pmatrix} = \begin{pmatrix} B_1 & 0 \\ 0 & * \end{pmatrix},$$

where E is a unit matrix of appropriate size.

(9.2) **Corollary.** *Let $b : N \times N \to \mathbb{Z}$ be a positive definite bilinear form with determinant 1. If $n = \dim(N) \leqslant 5$, then $(N,b) \cong (\mathbb{Z}^n, c)$, where c is the bilinear form given by the unit matrix. In other words, if B is a matrix belonging to b then $X \in \mathrm{GL}(n, \mathbb{Z})$ exists with*

$$XBX' = \begin{bmatrix} 1 & & 0 \\ & \ddots & \\ 0 & & 1 \end{bmatrix}.$$

PROOF. By Hermite's theorem, $e_1 \in N$ exists with $b(e_1, e_1) \leqslant (4/3)^2 < 2$. Hence $b(e_1, e_1) = 1$ because b assumes only integral values. Thus the one-dimensional submodule of N defined by e_1 is nonsingular. Hence, by the above lemma, there is an orthogonal decomposition $N = e_1\mathbb{Z} \perp N_1$, where (N_1, b) is unimodular also. Applying this procedure with N_1, one obtains the corollary.

Hermite showed that this latter result is true for $n \leqslant 7$; however it is not true for $n = 8$. We will return to this later.

Hermite's estimate has another important consequence which we already know from Chapter 4 in the case of two-dimensional forms.

(9.3) **Theorem** (Eisenstein, Hermite). *There are only finitely many isomorphism classes of positive definite (symmetric) bilinear spaces over $\mathbb{Z}$ of given dimension n and given determinant D.*

PROOF (By induction on n). The case $n = 1$ is clear. Let $B = (b_{ij})$ (B a positive definite symmetric matrix with coefficients in $\mathbb{Z}$ and) be the matrix of a positive definite bilinear space over $\mathbb{Z}$ with determinant D. Without loss of generality we can assume that b_{11} is minimal. Set

$$A = \begin{pmatrix} 1 & \alpha \\ 0 & E_{n-1} \end{pmatrix},$$

where $\alpha = (b_{21}/b_{11}, b_{31}/b_{11}, \ldots, b_{n1}/b_{11})$ and E_{n-1} is a unit matrix of rank $n - 1$. Then

$$B = A' \begin{bmatrix} b_{11} & 0 \\ 0 & \dfrac{1}{b_{11}} B' \end{bmatrix} A$$

with an integral symmetric matrix B' of rank $n - 1$ and $\det(B') = b_{11}^{n-2} D$. By (9.1) b_{11} can assume only finitely many values; this is, of course, true for the determinant of B' as well. Moreover, B' is positive definite. By induction, up to congruence there are only finitely many possibilities for B', say $B_1, \ldots, B_t$. Then there is a k, $1 \leqslant k \leqslant t$, and also an $X \in \mathrm{GL}(n - 1, \mathbb{Z})$ such that $B' = X' B_k X$, that is,

$$B = \begin{bmatrix} 1 & 0 \\ \alpha' & X' \end{bmatrix} \begin{bmatrix} b_{11} & 0 \\ 0 & \dfrac{1}{b_{11}} B_k \end{bmatrix} \begin{bmatrix} 1 & \alpha \\ 0 & X \end{bmatrix}.$$

Set

$$Y = \begin{pmatrix} 1 & 0 \\ 0 & X^{-1} \end{pmatrix}.$$

Then

$$Y' B Y = \begin{bmatrix} 1 & 0 \\ \beta' & E_{n-1} \end{bmatrix} \begin{bmatrix} b_{11} & 0 \\ 0 & \dfrac{1}{b_{11}} B_k \end{bmatrix} \begin{bmatrix} 1 & \beta \\ 0 & E_{n-1} \end{bmatrix}$$

with $\beta = \alpha X^{-1}$. Now we select a vector $u \in \mathbb{Z}^{n-1}$ such that the absolute value of each component $u + \beta = y$ does not exceed $\frac{1}{2}$. Set

$$Z = \begin{pmatrix} 1 & u \\ 0 & E_{n-1} \end{pmatrix};$$

then, with $Z' = YZ$,

$$\tilde{B} := Z'' B Z' = \begin{bmatrix} 1 & 0 \\ y' & E_{n-1} \end{bmatrix} \begin{bmatrix} b_{11} & 0 \\ 0 & \dfrac{1}{b_{11}} B_k \end{bmatrix} \begin{bmatrix} 1 & y \\ 0 & E_{n-1} \end{bmatrix}.$$

Obviously, there are only finitely many possibilities for $\tilde{B}$. The statement follows.

The further development of the theory of integral quadratic forms was strongly influenced by Hermann Minkowski. Minkowski was primarily a number theorist, but, motivated by his number-theoretical research, he became more and more interested in geometry and virtually created a new subject which he called the "geometry of numbers." It deals with the use of geometrical methods in number theory. Minkowski's first attempts in this direction probably occurred in 1889 when he studied Hermite's reduction theory. We do not know whether Minkowski attempted to simplify the proof of (9.1) or if he tried to improve the estimate for the minimum. In any event, on November 6, 1889, he wrote to Hilbert: "Perhaps you or Hurwitz are interested in the following theorem (which I can prove in half a page): in a positive quadratic form of determinant D with n ($\geqslant 2$) one always can assign such values to the variables that the form is $< nD^{1/n}$." Expressed in our terminology, this theorem says the following.

(9.4) Theorem. *Let* $b : \mathbb{Z}^n \times \mathbb{Z}^n \to \mathbb{Z}$ *be a positive definite symmetric bilinear form of determinant* D. *Then there exists* $x = (x_1, \ldots, x_n) \in \mathbb{Z}^n$, $x \neq 0$ *such that*

$$0 < b(x,x) < n\sqrt[n]{D} .$$

PROOF. In his proof, Minkowski identifies the given bilinear form with the usually Euclidean metric in $\mathbb{R}^n$, i.e., he performs a transformation $X = (\beta_{ij})$ such that

$$X'BX = \begin{bmatrix} 1 & & 0 \\ & \ddots & \\ 0 & & 1 \end{bmatrix}.$$

Here $B = (b_{ij})$, $b_{ij} = b(e_i, e_j)$ and $e_1, \ldots, e_n$ is the canonical basis of $\mathbb{R}^n$.

Under this transformation, the vectors $(x_1, \ldots, x_n) \in \mathbb{Z}^n$ correspond to the points of the lattice L in $\mathbb{R}^n$ formed by the column vectors of X^{-1}, i.e.,

$$L = \left\{ \sum_{i=1}^{n} \alpha_i b_i \mid \alpha_i \in \mathbb{Z} \right\}$$

for suitable vectors $b_1, \ldots, b_n$. We then have to find the minimal distance between different lattice points. Let E be the parallelotope with edges $b_1, \ldots, b_n$. We know that

$$\Delta = \Delta(L) := \text{vol}(E) = |\det(b_1, \ldots, b_n)| = |\det(X^{-1})|.$$

holds for the volume of E, the so-called *fundamental domain* (or *fundamental cell*) of L. (According to Weierstrass, this is virtually the definition of the determinant.) Instead of vol(E) we can write vol(L), since this number depends on L only. Since $\det(X^{-1}) = \sqrt{D}$, we also have $\Delta = \text{vol}(L) = \sqrt{D}$. The parallelotopes $E + x$, $x \in L$ fill the whole space $\mathbb{R}^n$ and have common edges. Around every lattice point as a center, Minkowski constructs an n-dimensional cube with edges of length $(1/\sqrt{n})M$, where M is the minimal distance between two different lattice points. Suppose all these cubes are parallel to each other, as shown in the figure below. According to the theorem of Pythagoras, the distance between the center of the cube to any of its corners is

$$\sqrt{n\left(\frac{1}{2}\frac{1}{\sqrt{n}}M\right)^2} = \frac{1}{2}M.$$

Since M is the minimal distance between two points of the lattice, all the cubes are disjoint. Then

$$\text{vol}(E) > \text{vol (cube)}, \qquad \text{that is} \quad \sqrt{D} > \left(\frac{1}{\sqrt{n}}M\right)^n$$

and hence

$$M^2 < n\sqrt[n]{D}.$$

Obviously, Minkowski's estimate is much better than Hermite's for large n. Minkowski immediately realized that his estimate can be improved once again, by using balls around every point of the lattice with radius $\frac{1}{2}M$. These balls touch each other, so that their volume is smaller than the volume of the fundamental domain, i.e.,

$$\left(\tfrac{1}{2}M\right)^n \omega_n < \sqrt{D},$$

where ω_n is the volume of the n-dimensional unit ball, i.e.,

$$\omega_1 = 2, \qquad \omega_2 = \pi, \qquad \omega_n = \omega_{n-2} \frac{2\pi}{n}.$$

One obtains:

(9.5) Theorem. *Let $b : \mathbb{Z}^n \times \mathbb{Z}^n \to \mathbb{Z}$ be a positive definite symmetric bilinear form of determinant D. Then $x = (x_1, \ldots, x_n) \in \mathbb{Z}^n$, $x \neq 0$, exists, such that*

$$b(x, x) \leqq 4 \sqrt[n]{D} \, \omega_n^{-2/n}.$$

The proof of this theorem is so simple and so natural that hardly any mathematician would have thought much about it. Minkowski did and analyzed what properties of the ball are really needed for the proof. If one asks this question it is easy to find the answer—one requires that the ball be symmetric with respect to its center and that its limiting surface, as Minkowski first expresses it, be nowhere concave. These properties ensure that the balls around the lattice points will all be disjoint. This leads to Minkowski's famous lattice point theorem which is the foundation of his "geometry of numbers." We formulate this theorem in the way it is usually applied.

(9.6) Minkowski's Lattice Point Theorem. *Let L be a lattice in $\mathbb{R}^n$ and K a centrally symmetric convex set around the origin, i.e., when $x, y \in K$, then $-x$ and $\frac{1}{2}(x + y) \in K$. Then, if $\mathrm{vol}(K) \geqslant 2^n \Delta(L)$, the set K contains a lattice point $x \in L$, $x \neq 0$.*

Let us give a brief sketch of the proof. First, let K be an arbitrary set with a well-defined volume, K disjoint from all the $K + x$, $x \in L$, $x \neq 0$. Then $\mathrm{vol}(K) \leqslant \mathrm{vol}(E)$, where E is a fundamental domain. Intuitively, this is obvious; one proves it by decomposing K in pieces $K_1, K_2, \ldots$, where the pieces lie in the different translates of the fundamental domain. Then one moves the pieces into a fixed fundamental domain where they are disjoint. This immediately gives our inequality (see sketch below). If $\mathrm{vol}(K) > 2^n \Delta$, i.e., $\mathrm{vol}(\frac{1}{2} K) > \Delta$ with $\frac{1}{2} K = \{ \frac{1}{2} x \mid x \in K \}$, then not all the parallel trans-

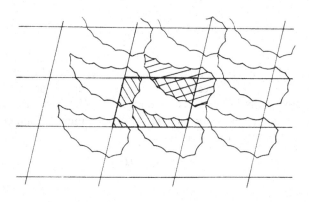

lates of $\frac{1}{2}K$ are disjoint. Therefore there are $\frac{1}{2}x, \frac{1}{2}y \in \frac{1}{2}K$ and $z \in L, z \neq 0$ with $\frac{1}{2}x = \frac{1}{2}y + z$ or $z = \frac{1}{2}(x - y)$. By our assumption, $-y$ and $\frac{1}{2}(x - y)$ lie in K, which completes the proof.

In his obituary of Minkowski, Hilbert made the following comment about this theorem and its proof.

> The proof of a deep number theoretical theorem without major computations, and basically with the help of geometric intuition is a pearl among Minkowski's inventions . . . . It is even more important that the basic idea of Minkowski's proof only uses the property of the ellipsoid that it is a convex figure with a center. Thus it can be transferred to arbitrary convex figures with center. This led Minkowski to see that the concept of convex bodies is a fundamental notion in our science, one of its most fertile tools for research.

Minkowski first published these ideas and theorems in his paper "Über positive quadratische Formen und über die kettenbruchähnlichen Algorithmen" which was published in Crelle's Journal (Volume 107, pp. 278–297) in 1891. He immediately saw that the lattice point theorem opened direct access to the proof of many fundamental facts of algebraic number theory. Examples are the theorem that the class number is finite and Dirichlet's unit theorem both of which are proved in modern texts on algebraic number theory using the lattice point theorem (see, for instance, Borevich and Shafarevich). Another important consequence of Minkowski's technique was a proof of Kronecker's conjecture that the discriminant of an algebraic number field $\neq \mathbb{Q}$ is always greater than 1, i.e., that at least one prime ramifies in such a field.

Since this text does not assume the basics of algebraic number theory, we will not continue our discussion of such topics. Instead, we will use the lattice point theorem to give new proofs of the following more elementary theorems.

Theorem (Fermat, Euler). *Every prime number p of the form for $4k + 1$ can be written as the sum of two squares of integers.*

Theorem (Lagrange, Euler). *Every natural number is the sum of four squares of integers.*

Theorem (Legendre). *Let a, b, c be relatively prime square-free integers which do not all have the same sign. The equation*

$$ax^2 + by^2 + cz^2 = 0$$

has a solution $(x, y, z) \neq (0, 0, 0)$ if and only if the following congruences are solvable:

$$u^2 \equiv -bc \pmod{a},$$

$$v^2 \equiv -ca \pmod{b},$$

$$w^2 \equiv -ab \pmod{c}.$$

One proves the first theorem in the following way. We have already mentioned that an integer u exists such that $u^2 \equiv -1 \mod p$. In $\mathbb{R}^2$ we consider the lattice

$$L = \{(x, y) \in \mathbb{Z}^2 \,|\, y \equiv ux \mod p\}.$$

Thus one can choose $x \in \mathbb{Z}$ arbitrarily. This x determines y up to multiples of p so that our lattice contains every pth integral pair x, y. The fundamental domain has area p. Setting $r = 2\sqrt{p/\pi}$ and $K = K(0, r)$, i.e., a disc around 0 with radius r, we obtain $\mathrm{vol}(K) \geq 2^2 \Delta(L)$. According to the lattice point theorem, $(x, y) \in L - \{0\}$ exists such that

$$0 < x^2 + y^2 \leq \frac{4p}{\pi} < 2p.$$

Since $u^2 \equiv -1 \mod p$ and $y \equiv ux \mod p$, we have $x^2 + y^2 \equiv 0 \mod p$ and $x^2 + y^2 = p$ follows.

Using the same technique, the reader can prove other theorems, among them the following two which go back to Euler:

Every prime number $p = 1 + 6k$ is of the form $x^2 + 3y^2$.

Every prime number $p = 1 + 8k$ is of the form $x^2 + 2y^2$.

To prove the four-square theorem, it suffices to show that every odd prime number p is the sum of four squares (see page 31).

We first mention that the equation $ax^2 + by^2 = c$ is solvable for all $a, b, c \in K$ in a finite field K. This is proved with the help of the so-called pigeonhole principle: let q be the number of elements in K. Without loss of generality we can assume that q is odd because the equation $ax^2 = c$ can be solved for even q. Substituting consecutively the elements of K for x provides $\frac{1}{2}(q + 1)$ different values for the expression ax^2. If one does the same for y, one obtains $\frac{1}{2}(q + 1)$ values for $c - by^2$. Since

$$\tfrac{1}{2}(q + 1) + \tfrac{1}{2}(q + 1) > q,$$

it follows that one of the values for ax^2 is identical to one of the values for $c - by^2$, i.e., $ax^2 + by^2 = c$ is solvable.

In our problem, this means that there are $u, v \in \mathbb{Z}$ with $u^2 + v^2 + 1 \equiv 0 \mod p$. Now we consider in $\mathbb{R}^4$ the lattice

$$L = \{(a, b, c, d) \,|\, c \equiv ua + vb \mod p,\ d \equiv ub - va \mod p\}$$

for which $\Delta(L) = p^2$. With $r = \sqrt[4]{32} \sqrt{p/\pi}$ one obtains $\mathrm{vol}(K(0, r)) = r^4 \omega_4 = 2^4 p^2 \geq 2^4 \Delta(L)$, and $(a, b, c, d) \in L - \{0\}$ with $0 < a^2 + b^2 + c^2 + d^2 < (4\sqrt{2}\,p)/\pi < 2p$ exists according to the lattice point theorem. Then $a^2 + b^2 + c^2 + d^2 = p$ because $a^2 + b^2 + c^2 + d^2 \equiv a^2 + b^2 + (ua + vb)^2 + (ub - va)^2 \mod p \equiv (a^2 + b^2)(u^2 + v^2 + 1) \mod p \equiv 0 \mod p$.

Concerning Legendre's theorem, it is easy to see that the given congruences are necessary. For $ax^2 + by^2 + cz^2 = 0$, one has $by^2 + cz^2 \equiv 0 \pmod{a}$ and consequently $(cz)^2 \equiv -bcy^2$. Since we can assume that x, y, z

are relatively prime, y is a unit $\bmod a$, consequently $x^2 = -bc \pmod{a}$ is solvable. Conversely, we consider the lattice L of all integral (x, y, z) with

$$uy \equiv cz \pmod{a},$$

$$vz \equiv ax \pmod{b},$$

$$wx \equiv by \pmod{c}$$

for a fixed solution (u, v, w) of the congruences in Legendre's theorem. It is easy to see that $\Delta(L) = |abc|$ and that these congruences lead to the congruence

$$ax^2 + by^2 + cz^2 \equiv 0 \pmod{abc}, \quad (x, y, z) \in L.$$

We know that the convex centrally symmetrical ellipsoid

$$K = \left\{ (x, y, z) \in \mathbb{R}^3 \,\middle|\, |a|x^2 + |b|y^2 + |c|z^2 \leqq R \right\}$$

has volume $(4\pi/3)(R^3/|abc|)^{1/2}$. According to the lattice point theorem, an element $(x, y, z) \in (L \cap K)$ with $(x, y, z) \neq 0$ exists if

$$\frac{4\pi}{3} \left(\frac{R^3}{|abc|} \right)^{1/2} > 8|abc|$$

or

$$R > \left(\frac{6}{\pi} \right)^{2/3} |abc|.$$

This means that $(x, y, z) \in L$ with $(x, y, z) \neq 0$ exists with

$$|ax^2 + by^2 + cz^2| \leqslant |a|x^2 + |b|y^2 + |c|z^2 < 2|abc|,$$

i.e., $ax^2 + by^2 + cz^2 = 0$ or $ax^2 + by^2 + cz^2 = \pm abc$. In the first case, we are finished. If $ax_0^2 + by_0^2 + cz_0^2 = -abc$, then

$$a(x_0 z_0 + by_0)^2 + b(y_0 z_0 - ax_0)^2 + c(z_0^2 + ab)^2 = 0,$$

and we are finished. To exclude the case $ax^2 + by^2 + cz^2 = abc$, we use the fact that a, b, c do not all have the same sign. We will not go into details here because we are only interested in the applications of the lattice point theorem.

Like nearly every other great mathematician, Minkowski was very interested in the applications of mathematics, specifically to physical questions. His five papers in physics made Minkowski's name known beyond the small circle of specialists. He was the first to formulate the basic equations of electrodynamics in a relativistic setting. Historically, it should be mentioned that essential parts of special relativity were formulated by Lorentz towards the end of the last century; 10 years later, Einstein developed them into the theory of special relativity. Einstein, incidentally, was a student of Minkowski when he was in Zürich but did not seem to understand much in his lectures. Central relativistic concepts, such as light cone, time and space vector go back to Minkowski; he postulated that gravitation propagates with the speed of light, something that has not been verified experimentally.

Having made these few remarks about Minkowski's work in physics, we turn to Minkowski's life. When Göttingen offered him a position as a full professor, he wrote:

Hermann Minkowski, born 22 June 1864 in Alexoten in Russia, attended the Altstädtische Gymnasium zu Königsberg in Preussen from 1872 to 1880. Starting on Easter 1880, he studied mathematics, first five semesters in Königsberg in Prussia with Heinrich Weber, then three semesters in Berlin with Kronecker and Weierstrass. He obtained his Ph.D. in Königsberg in Prussia on 30 July 1885 and obtained permission to teach at the University (Habilitation) in Bonn on 15 April 1887. He obtained a position as ausserordentlicher Professor in Bonn on 12 August 1892. In April of 1894, he moved to Königsberg and was made a full professor there on 18 March 1895. Minkowski resigned this position on 12 October 1896 to accept an offer as Professor of Mathematics at the Eidgenössische Polytechnikum in Zürich, a position which he held until the Fall of 1902. On 7 July 1902, Minkowski was appointed full Professor in Göttingen.

Minkowski had been a full Professor in Göttingen for seven years when he suddenly died of appendicitis on January 12, 1909.

We learn much about Minkowski's life, his person and his scientific work from a lengthy obituary by his close friend David Hilbert (1862–1943), the most important mathematician of the time and one of the most important mathematicians of all times, and the recollections of his daughter which were published as an introduction to an edition of his letters to Hilbert.

Charles Hermite

Hermann Minkowski

According to these sources, Minkowski was the scion of a family of Jewish merchants which came from the boundary area between Russia, Poland, Lithuania, and East Prussia (in terms of pre-World War I boundaries). The village Alexoten which is mentioned in his curriculum is located opposite the city of Kaunas (Kowno) on the River Memel. Political unrest, the Polish revolt against Russia, and discrimination against Jews at universities in Russia prompted the family to move to Königsberg, at that time the capital city of the German province of East Prussia. Like his brothers, Hermann Minkowski was extraordinarily gifted (his brother Oscar was an important medical researcher who, among other things, first understood the mechanism of diabetes) and, at a very early age, was able to obtain extraordinary scientific results. Hilbert writes:

Since he was very quick and had an excellent memory he finished secondary school very fast and was granted his abitur diploma in March 1880 when he was only 15 years old. On Easter 1880 Minkowski started to go to university. All together, he studied five semesters in Königsberg, mainly with Weber and Voigt, and three semesters in Berlin where he went to courses of lectures by Kummer, Kronecker, Weierstrass, Helmholtz and Kirchhoff. Very early on, Minkowski began his deep and thorough mathematical investigations. On Easter 1881, the Academy in Paris posed the problem of the composition of integers in sums of five squares as its prize essay. The seventeen-year-old student attacked this topic with all his energy and solved it brilliantly, developing, far beyond the original question, a general theory of quadratic forms, specifically their division in orders and genera, even for arbitrary rank. It is remarkable how well versed Minkowski was in algebraic methods, particularly the theory of elementary divisors, and in transcendental tools,

Gotthold Eisenstein

such as Dirichlet series and Gauss sums . . . . Not yet eighteen, Minkowski
submitted his paper to the Academy in Paris on 30 May 1882. Although it
was, contrary to the rules of the Academy, written in German, the Academy,
stressing the exceptional character of the case, gave Minkowski the full prize.
The report points out that a work of such importance should not be excluded
because of an irregularity in the form in which it was submitted. Minkowski
received the Grand Prix des Sciences Mathematiques in April 1883 . . . .
When it became known in Paris that he was supposed to obtain the prize of
the Academy, the nationalistic press started completely unfounded attacks
and rumors. The Academicians C. Jordan and J. Bertrand defended Minkowski immediately without any reservation: "Travaillez, je vous prie, à
devenir un géomètre éminent." This was the climax of the correspondence
between the great French mathematician C. Jordan and the young German
student—an admonition which Minkowski heeded. For him, a very productive period with many publications started.

The above estimates for the minimum M, by Hermite and Minkowski,
naturally lead to the question of the best possible bound, i.e., the question

of finding lattices, so-called extremal lattices, with determinant 1 and best possible M. Let μ_n be the largest value such that a lattice of determinant 1 exists in $\mathbb{R}^n$ for which the distances between different lattice points are all $\geqslant \mu_n$. This problem has only been solved for dimension $\leqslant 8$. One has:

(9.7) Theorem (Korkiné, Zolotareff, Blichfeldt). *For* $n = 2, \ldots, 8$, μ_n *assumes the values*

$$\sqrt[4]{4/3}\ ,\ \sqrt[6]{2}\ ,\ \sqrt[8]{4}\ ,\ \sqrt[10]{8}\ ,\ \sqrt[12]{64/3}\ ,\ \sqrt[14]{64}\ ,\ \sqrt{2}\ .$$

There is an easy proof only when $n = 2$. In this case, the lattice is constructed by covering the plane with isosceles triangles; in other words, it

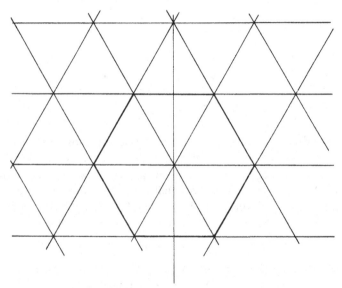

is spanned by the vectors $(\sqrt[4]{4/3}\,, 0)$ and $(\frac{1}{2}\sqrt[4]{4/3}\,, \sqrt[4]{3/4}\,)$. The fundamental domain obviously has area $1 = \sqrt[4]{4/3}\ \cdot\ \sqrt[4]{3/4}$; one can easily show that M is maximal for this lattice. (Proof: Consider an extremal lattice. Without loss of generality, let $(a, 0)$, $a > 0$ be a lattice point with minimal distance from 0. Then the second basis vector can be written in the form $(b, 1/a)$. Let $\mathbf{c}$ be parallel to the axis of the lattice with distance $1/a$; then $\mathbf{c}$ contains lattice points with the distance a. Choosing the point closest to the y-axis, we can ensure that $b \leqslant a/2$. If $a > \sqrt[4]{4/3}$, we would have $a^4 > \frac{4}{3}$ and consequently $1/a^2 < 3a^2/4$, i.e., $a^2/4 + 1/a^2 < a^2$, and $\sqrt{a^2/4 + 1/a^2} < a$ and $\sqrt{b^2 + 1/a^2} < a$. This says that $(b, 1/a)$ has a smaller distance to the origin than a, a contradiction to our assumption.)

We will now describe the lattices for which the maximum μ_n is assumed, in dimensions 2 to 8. First we consider the following graphs [these graphs are important in many areas of mathematics, specifically in the theory of

Lie groups (whence the notation $A_2, \ldots, E_8$) or in the determination of all Platonic solids]. For any of these graphs with n points, one constructs a lattice in $\mathbb{R}^n$ in the following way. A basis vector of the lattice corresponds to each point of the graph. All basis vectors have the same length. If there is a direct link between the points that correspond to two basis vectors, the basis vectors are connected by an angle of 60°; otherwise, they are orthogonal to each other. One finally determines the length of the basis vectors in such a way that the fundamental cell has volume 1. This determines the lattice.

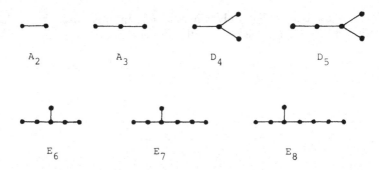

We have already mentioned that it is difficult to show that these lattices have the required property, but it is easy to calculate that μ_n has the given values. Let $b_1, \ldots, b_n$ be basis vectors which, for now, are all of length $\sqrt{2}$, i.e., their inner product $\langle b_i, b_i \rangle$ is equal to 2. The inner product between two neighboring basis vectors is 1 ($= 2 \cos 60°$), and for two basis vectors that are not neighbors, the inner product is 0. In other words,

$$\langle b_i, b_j \rangle = \begin{cases} 2, & b_i = b_j, \\ 1, & b_i, b_j \text{ neighbors}, \\ 0, & b_i, b_j \text{ not neighbors}. \end{cases}$$

$$\begin{pmatrix} 2 & 1 \\ 1 & 2 \end{pmatrix}, \begin{bmatrix} 2 & 1 & 0 \\ 1 & 2 & 1 \\ 0 & 1 & 2 \end{bmatrix}, \ldots, \begin{bmatrix} 2 & 1 & 0 & 0 & 0 & 0 & 0 & 0 \\ 1 & 2 & 1 & 0 & 0 & 0 & 0 & 0 \\ 0 & 1 & 2 & 1 & 1 & 0 & 0 & 0 \\ 0 & 0 & 1 & 2 & 0 & 0 & 0 & 0 \\ 0 & 0 & 1 & 0 & 2 & 1 & 0 & 0 \\ 0 & 0 & 0 & 0 & 0 & 2 & 1 & 0 \\ 0 & 0 & 0 & 0 & 0 & 1 & 2 & 1 \\ 0 & 0 & 0 & 0 & 0 & 0 & 1 & 2 \end{bmatrix}$$

are the corresponding matrices, and it is not difficult to show that they are positive definite.

It is well known that the determinant of the matrix $B = (\langle b_i, b_j \rangle)$ is equal to the square of the volume of the fundamental domain. (Write b_i as a linear combination of the orthonormal basis $e_1, \ldots, e_n$, $b_i = \sum \beta_{ij} e_j$. Then $B = (\beta_{ij})'(\beta_{ij})$.)

None of the latice points are closer to the origin than b_i; for suppose $0 \neq \sum_{i=1}^{n} k_i b_i$, $k_i \in \mathbb{Z}$ is an arbitrary lattice point; then

$$\left\langle \sum_{i=1}^{n} k_i b_i, \sum_{j=1}^{n} k_j b_j \right\rangle = \sum \sum k_i k_j \langle b_i, b_j \rangle$$

$$= \sum_{i=1}^{n} k_i^2 \cdot 2 + 2 \cdot \sum_{\substack{i,j \text{ adjacent} \\ i<j}} k_i k_j \geq 2.$$

One obtains the number μ_n by explicitly computing $\det(B)$. Let us look at the case $n = 3$ as an example. Then

$$B = \begin{bmatrix} 2 & 1 & 0 \\ 1 & 2 & 1 \\ 0 & 1 & 2 \end{bmatrix}, \qquad \det(B) = 4.$$

This means that the volume of the fundamental cell is 2. To obtain the volume 1, one has to multiply everything by the factor $1/\sqrt[3]{2}$. The minimal distance is $\sqrt{2}/\sqrt[3]{2} = \sqrt[6]{2}$. This lattice occurs in the crystal structure of many elements, among them silver and gold. In general, if one wants to obtain the volume 1 for the fundamental cell, one has to multiply the basis vectors by the factor $\alpha = 1/\sqrt[2n]{\det B}$. The minimal distance is $\alpha\sqrt{2}$.

The case $n = 8$ is particularly interesting. Then $\det B = 1$ and the minimal distance is $\sqrt{2}$. The corresponding symmetric bilinear form b is not isomorphic to the unit form $x_1^2 + \cdots + x_8^2$ because b assumes only even values.

The question of the maximum μ_n is connected to the problem of densest *sphere packings*. This is the problem of packing infinitely many spheres (of equal radius) in $\mathbb{R}^n$ as densely as possible. This problem is open for all $n \geq 3$ even though all physicists "know" (and no mathematician would doubt this) that the lattice that belongs to A_3 supplies the densest packing in $\mathbb{R}^3$; every lattice point will be the center of a sphere.

To conclude this topic we want to draw the reader's attention to another unsolved problem which is of practical importance in the theory of crystal structures. If one tries to determine the crystal structure of an object, one measures the distances between the different lattice planes by x-ray interference (the Debye–Scherrer experiment). We do not know whether these distances determine the lattice uniquely.

References

Ch. Hermite: *Oeuvres*, 4 Bände, Gauthier-Villars, Paris, 1905. Specifically: Lettres de M. Hermite à M. Jacobi sur différents objects de la théorie de nombres.

H. Minkowski: *Gesammelte Abhandlungen*. Specifically: Über die positiven quadratischen Formen und über kettenbruchähnliche Algorithmen, Bd. II, 243–260. Zur Theorie der positiven quadratischen Formen, Bd. II, 212–218. Gedächtnisrede auf H. Minkowski, von D. Hilbert, Seite V–XXXI.

H. Minkowski: *Geometrie der Zahlen*, Teubner, Leipzig, 1896. Johnson Reprint Corp., New York 1968.

H. F. Blichfeldt: The minimum values of positive quadratic forms in six, seven and eight variables, Mathematische Zeitschrift, **39**, 1934, 1–15.

A. N. Korkine, E. I. Zolotarev: Sur les formes quadratiques positives quaternaires, Mathematische Annalen, **5**, 1872, 581–583. Sur les formes quadratiques, Mathematische Annalen, **6**, 1873, 366–389. Sur les formes quadratiques positives, Mathematische Annalen, **11**, 1877, 242–292.

C. A. Rogers: *Packing and Covering*, Cambridge University Press, 1964.

J. Milnor, D. Husemoller: *Symmetric Bilinear Forms*, Springer-Verlag, 1973.

H. Davenport, M. Hall: On the Equation $ax^2 + by^2 + cz^2 = 0$, Quart. J. Math. (2) **19**, 1948, 189–192.

CHAPTER 10

Preview: Reduction Theory

The main emphasis of this book has been on the theory of quadratic forms, and we have given special attention to reduction theory. The main question of reduction theory can be formulated in the following way. Let us consider the real-valued quadratic forms in n variables. We look for inequalities for the coefficients such that every form is integrally equivalent to one and only one reduced form, i.e., to a form which satisfies all these inequalities. (From now on, without again stating this explicitly, we will confine ourselves to positive forms.)

In the simplest case of forms of two variables, we have described the solution of this problem by Lagrange and discussed some of its applications. The question is much more complicated for three variables. It was solved by Seeber, a student of Gauss. Dirichlet gave a proof using only methods from elementary geometry (see References). If the form is given by the matrix

$$\begin{bmatrix} a & d & e \\ d & b & f \\ e & f & c \end{bmatrix},$$

the Seeber–Dirichlet conditions are the following:

$$0 < a \leqq b \leqq c,$$

$$2|d| \leqq a, \qquad 2|e| \leqq a, \qquad 2|f| \leqq b,$$

either $\quad d, e, f \geqq 0$

or $\qquad d, e, f \leqq 0 \quad$ and $\quad -2(d + e + f) \leqq a + b.$

(There are additional constraints in exceptional cases such as $a = b$, etc., which we will not discuss.)

For more than three variables, the problem is so complicated that an explicit solution is not known (and would probably not be of any interest because it would be so complicated). Minkowski, in one of his most important papers, gave a nonconstructive solution to the problem. This we will now describe. A quadratic form in n variables or a symmetric $n \times n$ matrix has $\frac{1}{2}n(n+1) = N$ free coefficients and will thus be interpreted as an element of the Euclidean space $\mathbb{R}^N$. Then:

(10.1) Theorem (Minkowski). *There is a convex cone H in $\mathbb{R}^n$, with finitely many hyperplanes through the origin as boundaries, such that for every positive symmetric matrix $A \in \mathbb{R}^n$, an integrally equivalent matrix $TAT^t \in H$ exists. It is uniquely determined if it is in the interior of H. (H contains positive forms only.)*

The theorem states that a general reduction theory similar to the theory for two and three variables exists. Minkowski proves a second basic theorem by considering the part H_1 of H which contains all symmetric matrices of determinant $\leqslant 1$. He calculates the volume of H_1:

(10.2) Theorem (Minkowski).

$$\text{volume}(H_1) = \frac{2^n}{n+1} \frac{\zeta(2) \dots \zeta(n)}{s_2 \dots s_n},$$

where s_k is the surface area of the k-dimensional unit sphere, i.e.,

$$s_k = \frac{2\pi^{k/2}}{\Gamma(k/2)} = \begin{cases} \dfrac{2\pi^{m+1}}{m!} & \text{for} \quad k = 2(m+1), \\[2mm] \dfrac{2^{2m+1}\pi^m m!}{(2m)!} & \text{for} \quad k = 2m+1. \end{cases}$$

First note that H_1 is not uniquely determined. However, since a change of variables with determinant 1 does not change the volume, $\text{vol}(H_1)$ is uniquely determined.

Perhaps the most interesting aspect of this deep and difficult theorem is that these formulas inductively yield geometrical interpretations for all values $\zeta(n)$ (also for odd n, see the chapter on Euler). Look, for example, at the case $n = 2$. Then $\text{vol}(H_1) = 2\zeta(2)/3\pi$; on the other hand, using Lagrange's explicit description of H_1, one obtains $\text{vol}(H_1) = \pi/9$ after a computation of approximately half a page length. This is equivalent to Euler's formula

$$\zeta(2) = \frac{\pi^2}{6}.$$

Analogously, for $n = 3$ one obtains

$$\zeta(3) = 24\,\text{vol}(H_1).$$

The Seeber–Dirichlet inequalities and the condition on the determinant describe H_1 explicitly, which means that this formula can be seen as a satisfactory interpretation of $\zeta(3)$. (Alternatively, one might attempt to calculate the sextuple integral for $\mathrm{vol}(H_1)$, but presumably no one has ever attempted this.)

Minkowski had a completely different reason for calculating the volume of H_1 (one should not forget that there was no way to guess what the result would be). According to Hermite (9.3), the class number $h(d)$ is finite for the forms of determinant d. The volume of H_1 can be used to describe the asymptotic growth $h(d)$. For a positive real number t, the area of tH_1 contains all reduced forms with determinant $\leq t^n$. This area has volume $t^N \mathrm{vol}(H_1)$, which means that the volume of the space of all reduced forms with determinants $\leq d$ equals $d^{(n+1)/2}\mathrm{vol}(H_1)$. For large d, this volume is approximately equal to the number of integral lattice points (corresponding to the integral quadratic forms) in this area, and we obtain

$$\lim_{d\to\infty} \frac{h(1) + \cdots + h(d)}{d^{(n+1)/2}} = \mathrm{vol}(H_1). \tag{10.3}$$

[If one were to assume that $h(d)$ grows uniformly (which is not the case), differentiating would yield that $h(d)$ grows like $d^{(n-1)/2}$ up to a constant factor. For the case $n = 2$, C. L. Siegel has shown that the class numbers of the imaginary quadratic fields with discriminant d satisfy the asymptotic law $\log h(d) \sim \log(\sqrt{d}\,)$.]

For $n = 2$, Gauss found a formula of the type (10.3), by a "theoretical investigation," as he expresses it in §302 of *Disquisitiones*. He did not prove this but one would suspect that he actually did have a proof because he would certainly have mentioned "induction" if he had relied on numerical evidence only. Altogether, §§301–303 contain a number of interesting asymptotic formulas and statements on classes and genera which even today have not all been proved. In fact, the situation which Gauss considers is somewhat more complicated because he only deals with primitive forms, i.e., forms whose coefficients do not have a common divisor. This leads to a factor $\zeta(3)$, which is explained by the following observation of Gauss: When one considers n-tuples of integers, $\zeta(n)^{-1}$ is the probability that these n numbers are relatively prime because $1 - 1/p^n$ is the probability that they do not all have the factor p for any prime number p. This means that

$$\prod_p (1 - p^{-n}) = \zeta(n)^{-1}$$

describes the probability that these numbers do not have any prime number as a common factor.

Minkowski published Theorems (10.1), (10.2), and (10.3) in his paper "Diskontinuitätsbereich für arithmetische Äquivalenz" in 1905 in Crelle's journal in an anniversary volume on the occasion of Dirichlet's one-hundredth birthday. The paper begins with the following characteristic sentence. "This paper repeatedly uses methods which were developed by Dirichlet." This clearly describes the tradition in which Minkowski saw himself and his mathematical research.

It was clear to Minkowski that the reduction theory of quadratic forms has a natural place in a more comprehensive "arithmetical theory of the group of all linear transformations." However, he did not expand this theory, but the program was picked up by C. L. Siegel and A. Weil and today occupies a central place in mathematical research (computation of Tamagawa numbers, etc.). We use the example of $SL(n, \mathbb{R})$ to describe these efforts. In the course of our computations, we will use analytical tools which were not yet available to Dirichlet and Minkowski, i.e., integration on locally compact topological groups. (A systematic presentation of algebraic number theory with thorough use of integration in topological groups is given in A. Weil's *Basic Number Theory*, Springer-Verlag, 1967.)

Let $G = SL(n, \mathbb{R})$ and $\Gamma = SL(n, \mathbb{Z})$. We investigate the homogeneous space G/Γ. Let dg be the Haar measure in G which induces the metric $\|X\|^2 = \mathrm{trace}(X'X)$ on the Lie algebra $sl(n, \mathbb{R})$. Let us now prove the main result of this chapter.

(10.4) **Theorem.** G/Γ *has the finite volume* $\zeta(2) \ldots \zeta(n)\sqrt{n}$.

PROOF. The proof is by induction on n (for $n = 1$ trivially $\mathrm{vol}(G/\Gamma) = 1$). Let us consider the "parabolic subgroup"

$$P = \left\{ \begin{pmatrix} 1 & x \\ 0 & g' \end{pmatrix} \Big| \, x \in \mathbb{R}^{n-1}, g' \in G' \right\}.$$

(A prime $'$ denotes corresponding objects for $n - 1$, i.e., $G' = SL(n - 1, \mathbb{R})$ etc.) P is the isotropy group of the first basis vector e_1. Futhermore,

$$T = \{(t_1, \ldots, t_n) \mid t_i \in \mathbb{Z}, \text{g.c.d.}(t_1, \ldots, t_n) = 1\}.$$

Γ operates transitively on T. Then one obtains for arbitrary integrable functions $\phi : \mathbb{R}^n \to \mathbb{R}_+$ with compact support,

$$\int_{G/\Gamma} \left(\sum_{t \in T} \phi(gt) \right) dg = \int_{G/\Gamma} \left(\sum_{\gamma \in \Gamma/\Gamma \cap P} \phi(g\gamma e_1) \right) dg$$

$$= \int_{G/\Gamma \cap P} \phi(ge_1) \, dg.$$

We now come to the decisive step in the proof: dg induces measures $d\bar{g}$ and dp on G/P and $P/\Gamma \cap P$ such that we can use Fubini's theorem and write the latter integral as the double integral,

$$\int_{G/P} \left(\int_{P/\Gamma \cap P} \phi(\bar{g}pe_1) \, dp \right) d\bar{g} = \int_{G/P} \phi(\bar{g}e_1) \, d\bar{g} \int_{P/\Gamma \cap P} dp$$

$$= \mathrm{vol}(P/\Gamma \cap P) \int_{G/P} \phi(\bar{g}e_1) \, d\bar{g}.$$

P is the semidirect product of $G' \cong \{(\begin{smallmatrix} 1 & 0 \\ 0 & g \end{smallmatrix})\}$ and $\mathbb{R}^{n-1} \cong \{(\begin{smallmatrix} 1 & x \\ 0 & 1 \end{smallmatrix})\}$. Since G' operates on $\mathbb{R}^{n-1}$ isometrically, dg yields the product measure $dg'\, dx$, where dx is the Euclidean measure. Analogously, $\Gamma \cap P$ is the semidirect product of Γ' and $\mathbb{R}^{n-1} \cap \Gamma = \mathbb{Z}^{n-1}$, hence

$$\mathrm{vol}\left(\frac{P}{\Gamma \cap P} \right) = \mathrm{vol}\left(\frac{\mathbb{R}^{n-1}}{\mathbb{Z}^{n-1}} \right) \mathrm{vol}\left(\frac{G'}{\Gamma'} \right) = \mathrm{vol}\left(\frac{G'}{\Gamma'} \right).$$

Furthermore, the natural operation of G on $\mathbb{R}^n - 0$ introduces a homomorphism $G/P \cong \mathbb{R}^n - 0$. The orthogonal space with respect to the quadratic form $\mathrm{trace}(X'X)$ of the Lie algebra of P is formed by the matrices

$$\begin{bmatrix} x_1 & & & \\ x_2 & y & & \\ \vdots & & \ddots & \\ x_n & & & y \end{bmatrix}, \qquad y = -\frac{x_1}{n-1}.$$

Then

$$\sqrt{\frac{n}{n-1}}\; dx_1 \ldots dx_n$$

is the induced measure.

We immediately obtain the formula

$$\sqrt{\frac{n}{n-1}}\; \mathrm{vol}(G'/\Gamma') \int_{\mathbb{R}^n} \phi(x)\, dx = \int_{G/\Gamma} \left(\sum_{t \in T} \phi(gt) \right) dg.$$

Setting $C = \sqrt{n/(n-1)}\; \mathrm{vol}(G'/\Gamma')$ and multiplying by $\zeta(n)$, we obtain

$$C\zeta(n) \int_{\mathbb{R}^n} \phi(x)\, dx = C \sum_{i=1}^{\infty} \int_{\mathbb{R}^n} i^{-n} \phi(x)\, dx$$

$$= C \sum_{i=1}^{\infty} \int_{\mathbb{R}^n} \phi(ix)\, dx = \sum_{i=1}^{\infty} \int_{G/\Gamma} \left(\sum_{t \in T} \phi(git) \right) dg$$

$$= \int_{G/\Gamma} \left(\sum_{i=1}^{\infty} \sum_{t \in T} \phi(git) \right) dg = \int_{G/\Gamma} \left(\sum_{\substack{u \in \mathbb{Z}^n \\ u \neq 0}} \phi(gu) \right) dg.$$

For the last equality, we have used the fact that every $u \in \mathbb{Z}^n$ can uniquely be written with $t \in T$. With ϕ the characteristic function χ_K of a compact set K, we obtain the formula

$$C\zeta(n)\mathrm{vol}(K) = \int_{G/\Gamma} \sum_{\substack{u \in \mathbb{Z}^n \\ u \neq 0}} \chi_K(gu)\, dg.$$

Let W_τ be the cube with $|x_i| \leq \tau$ and $h(g,\tau)$ be the number of u with $gu \in W_\tau$, i.e.,

$$h(g,\tau) = \sum_{\substack{u \in \mathbb{Z}^n \\ u \neq 0}} \chi_{W\tau}(gu).$$

Then $gu \in W_\tau$ if and only if $u \in g^{-1}W_\tau$. We now use Gauss' and Dirichlet's technique to compute the volume by estimating the number of lattice points. Then

$$\lim_{\tau \to \infty} \tau^{-n} h(g,\tau) = \mathrm{vol}(g^{-1}W_1) = 2^n.$$

This immediately yields

$$2^n \mathrm{vol}(G/\Gamma) = \int_{G/\Gamma} \lim_{\tau \to \infty} (\tau^{-n} h(g,\tau)) \, dg$$

$$= \lim_{\tau \to \infty} \tau^{-n} \int_{G/\Gamma} \left(\sum \chi_{W_\tau}(gu) \, dg \right) = \zeta(n) C 2^n,$$

that is, $\mathrm{vol}(G/\Gamma) = \sqrt{n/(n-1)} \; \zeta(n) \mathrm{vol}(G'/\Gamma')$.

Unfortunately this last step is not justified. Offhand, we can not exchange integral and limit. Instead of the equality, one only has $\leqslant$, i.e., at least the finiteness of the volume $\mathrm{vol}(G/\Gamma)$. Weil overcomes this problem by the technique which Dirichlet uses to compute the Gaussian sums and which we have occasion to admire here once more.

Let $\hat{\phi}(y)$ be the Fourier transform of $\phi(x)$, i.e.,

$$\hat{\phi}(y) = \int_{\mathbb{R}^n} \phi(x) e^{-2\pi i \langle x,y \rangle} \, dx,$$

and specifically

$$\hat{\phi}(0) = \int_{\mathbb{R}^n} \phi(x) \, dx.$$

We can use the Poisson summation formula

$$\sum_{u \in \mathbb{Z}^n} \phi(gu) = \sum_{v \in \mathbb{Z}^n} \hat{\phi}(g^{-1}v).$$

(One proves this by expanding the periodic function $\sum_{u \in \mathbb{Z}^n} \phi(g(u+x))$ into a uniformly convergent Fourier series and computing the expansion at $x = 0$.) Then we obtain

$$C\zeta(n) \int_{\mathbb{R}^n} \phi(x) \, dx + \phi(0) \mathrm{vol}(G/\Gamma)$$

$$= \int_{G/\Gamma} \sum_{u \in \mathbb{Z}^n} \phi(gu) \, dg$$

$$= \int_{G/\Gamma} \sum_{v \in \mathbb{Z}^n} \hat{\phi}(g^{-1}v) \, dg = \int_{G/\Gamma} \sum_{v \in \mathbb{Z}^n} \hat{\phi}(gv) \, dg$$

$$= C\zeta(n) \int_{\mathbb{R}^n} \hat{\phi}(x) \, dx + \hat{\phi}(0) \mathrm{vol}(G/\Gamma),$$

i.e.,

$$(C\zeta(n) - \mathrm{vol}(G/\Gamma))(\hat{\phi}(0) - \phi(0)).$$

Since this is true for every function ϕ, we miraculously obtain

$$\text{vol}(G/\Gamma) = \zeta(n)\sqrt{\frac{n}{n-1}}\ \text{vol}(G'/\Gamma').$$

This completes the proof of Theorem (10.4). We did not establish the connection to quadratic forms, but we at least saw the natural geometrical interpretation of the integral ζ-values. This was the main objective of this chapter. Let us now give a quick sketch of the connection with quadratic forms. Let P be the set of positive symmetric $(n \times n)$ matrices. P can be interpreted as an open subset of $\mathbb{R}^N$, $N = n(n+1)/2$. Our further argument is based on a study of the mapping

$$\psi : G \to P, \qquad g \to g'g.$$

Obviously, $g'g$ is symmetric and positive for every invertible matrix. We know from linear algebra (Sylvester's law of inertia) that every $q \in P$ can be written in this form. Then the mapping ψ is surjective. The "fibers" of ψ are the cosets of the orthogonal groups $0(n) = \{ g \in G \,|\, g'g = e \}$, since $0(n)g$ is obviously the pre-image of $g'g$.

With the help of the mapping ψ one can easily show that it follows from (10.1) that a fundamental domain F exists for $\text{GL}(n, \mathbb{R})/\text{GL}(n, \mathbb{Z})$: there is a "reasonable" set F with $\text{GL}(n, \mathbb{R}) = \bigcup_{\gamma \in \text{GL}(n,\mathbb{Z})} \gamma F$ such that F and γF do not have any common interior points for $\gamma \neq 1$. For if one sets

$$F = \left\{ g \in G \,|\, g'g \in H, \begin{array}{l} \det g \geqslant 0 \quad \text{if } n \text{ odd} \\ \text{trace } g \geqslant 0 \text{ if } n \text{ even} \end{array} \right\},$$

one obtains a fundamental domain F_1 for $\text{SL}(n, \mathbb{R})/\text{SL}(n, \mathbb{Z}) = G/\Gamma$, namely

$$F_1 = \begin{cases} F \cap G & \text{if } n \text{ odd,} \\[2mm] (F \cup g_0 F) \cap G & \text{if } n \text{ even,} \quad g_0 = \begin{bmatrix} -1 & & & \\ & 1 & & \\ & & 1 & \\ & & & \ddots \end{bmatrix}. \end{cases}$$

We leave the necessary easy computations to the reader. We computed the volume of F_1 (with respect to Haar measure) in (10.4). Due to the condition $\det(g) > 0$ (or trace$(g) \geqslant 0$) in the definition of F, the set $\psi^{-1}(H_1)$ consists of two parts. They have the same volume as the cone K over F_1 with the vertex at the origin. Comparing the Haar measure with the Euclidean measure $d\mu$ shows that the cone has Euclidean measure $(1/n)\zeta(2) \ldots \zeta(n)$, and $\psi^{-1}(H_1)$ the volume $(2/n)\zeta(2) \ldots \zeta(n)$. Now it is possible to compute the volume of H_1 with the help of the following transformation formula. In a way, it is a combination of the theorems of Fubini and the formula for the changing of variables. Let M be measurable and $f: P \to \mathbb{R}$ integrable. Then $f(\psi)$ is integrable and

$$\int_{\psi^{-1}(M)} f(\psi(g)) \, d\mu(g) = \frac{s_2 \cdots s_n}{2^{n-1}} \int_M \det(a)^{-1/2} f(a) \, d\mu(a). \quad (10.5)$$

PROOF OF (10.2). Let $f(a) = \det(a)^{1/2}$. With the help of (10.5) one obtains

$$\mu(H_1) = \int_{H_1} d\mu(a) = \int_{H_1} \det(a)^{-1/2} \det(a)^{1/2} \, d\mu(a)$$

$$= \frac{2^{n-1}}{s_2 \cdots s_n} \int_{\psi^{-1}(H_1)} |\det g| \, d\mu(g).$$

For odd n, $\det(g) > 0$ and

$$\mu(H_1) = \frac{2^n}{s_2 \cdots s_n} \int_K \det g \, d\mu(g).$$

On the cone K, one has, for any integral of a function depending only on $\det(g)$,

$$\int_K h(\det g) \, d\mu(g) = \mathrm{vol}(K) \int_0^1 h(t^{1/n}) \, dt. \quad (10.6)$$

The statement

$$\mu(H_1) = \frac{2^n}{s_2 \cdots s_n} \frac{\zeta(2) \cdots \zeta(n)}{n} \frac{n}{n+1}$$

follows immediately.

It is necessary to prove (10.6) only for the characteristic function χ of $(0, c)$; once proved for χ, one can generalize it to a step function and so on. One has

$$\int_K \chi(\det g) \, d\mu(g) = \mathrm{vol}\{ g \in K \mid \det(g) < c \}$$

$$= \mathrm{vol}(c^{1/n} K) = c^n \mathrm{vol}(K) = \mathrm{vol}(K) \int_0^1 \chi(t^{1/n}) \, dt.$$

Formula (10.5) has nothing to do with number theory; it is a statement from the integration theory of several variables, which we will not prove here, although we have to admit that there does not seem to be an elementary proof in the literature.

References

H. Minkowski: *Gesammelte Abhandlungen*. (Particularly: Diskontinuitätsbereich für arithmetische Äquivalenz, Band 2, S. 53–100.)

C. L. Siegel: *Gesammelte Abhandlungen*, 4 vols., Springer-Verlag, Berlin, Heidelberg, New York, 1966, 1979. (Particularly: *The Volume of the Fundamental Domain for Some Infinite Groups*, Vol. 1, 459–468; *A Mean Value Theorem in the Geometry of Numbers*, Vol. 3, 39–46.)

A. Weil: *Collected Papers*, 3 Vols., Springer-Verlag, 1979, Berlin, Heidelberg, New York. (Particularly: *Sur quelques résultats de Siegel*, Vol. 1, 339–357.)

G. P. L. Dirichlet: Über die Reduktion der positiven quadratischen Formen mit drei unbestimmten ganzen Zahlen. Werke II.

APPENDIX
English Translation of Gauss's Letter to Dirichlet, November 1838

The following is a translation of the letter on pp. 97–100:

Göttingen, November 2, 1838

I owe you, my most esteemed friend, my thanks for sending me your beautiful papers as well as for the kind words which you sent along with them. However, I am sorry that my expectations to see you here came to naught this time—I am very disappointed since to be with you during this dark period would have cheered me up, too.

You mentioned my earlier communication to Mr. Krone and the discretion I had asked him for. I wish you might not overinterpret the letter and explain to you that, by telling him right away that he should not make anything public, either directly or indirectly by passing it along to others, I wanted to retain the possibility to publish my investigations myself; as soon as I am no longer interested in working out my results, this possibility is no longer of any concern. I would be very pleased to explain the subject to you, but two circumstances would have to come together: from your end, a somewhat extended stay here, and from mine, sufficient leisure (and cheerfulness) to bring the subjects into the order necessary for further communication. This is the more difficult since little, and nothing organized, has been written down by me. However, I can assure you of my desire that circumstances will permit me to work things out.

However, in the immediate future I will have the little time that is left to me from business that I must call nonscientific, to complete another investigation which meanwhile might not be without interest to you.

When you mentioned in your letter several subjects in higher arithmetic, my heart fills with pain. For the higher I place this part of mathematics above all others (and have always done so), the more painful it is to me that I am so far removed from my favorite occupation, directly or indirectly by the circumstances. Again and again I had to delay my theory of the number of classes of quadratic forms of which I was already in possession in 1801 and to whose completion I have looked forward as a particularly pleasant job. Two or three years ago I thought to have found the time and have indeed made some progress; the opportunity presented several new and interesting results, however not in relation to the theory itself, which is complete since 1801, but rather in relation to insights that lead into it. However, I had to interrupt myself again and have not been able to resume my work, painful as this was. I am sure you yourself know from your own experience what such a resumption would mean. This is not like an everyday job which one can interrupt any minute and resume again. One always has to invest a lot of effort and has to have much free time to again bring everything to one's attention.

I was not aware of the quarrels of Poinsot or Poisson over the attraction of elliptical spheroids of which you wrote me. However, going through the Com[p]tes Rendus, I realized that such a quarrel was mentioned but I did skip those pages because I am disgusted by such arguments. Indeed, just as I like nothing better than when I realize that someone is interested in science for the sake of science, in the same way nothing is more disgusting to me than persons whom, for their talents, I hold in high regard show the pettiness of their characters.

Our beloved Weber sends his cordial regards and I recommend myself to your kind memory.

Yours, C. F. Gauss

Index

Undergraduate Texts in Mathematics

continued from ii

Malitz: Introduction to Mathematical
Logic: Set Theory - Computable
Functions - Model Theory.
1979. xii, 198 pages. 2 illus.

Martin: The Foundations of Geometry
and the Non-Euclidean Plane.
1975. xvi, 509 pages. 263 illus.

Martin: Transformation Geometry: An
Introduction to Symmetry.
1982. xii, 237 pages. 209 illus.

Millman/Parker: Geometry: A Metric
Approach with Models.
1981. viii, 355 pages. 259 illus.

Owen: A First Course in the
Mathematical Foundations of
Thermodynamics.
1984. xvii, 178 pages. 52 illus.

Prenowitz/Jantosciak: Join Geometrics:
A Theory of Convex Set and Linear
Geometry.
1979. xxii, 534 pages. 404 illus.

Priestly: Calculus: An Historical
Approach.
1979. xvii, 448 pages. 335 illus.

Protter/Morrey: A First Course in Real
Analysis.
1977. xii, 507 pages. 135 illus.

Ross: Elementary Analysis: The Theory
of Calculus.
1980. viii, 264 pages. 34 illus.

Scharlau/Opolka: From Fermat to
Minkowski: Lectures on the Theory of
Numbers and Its Historical Development.
1984. xi, 179 pages. 28 illus.

Sigler: Algebra.
1976. xii, 419 pages. 27 illus.

Simmonds: A Brief on Tensor
Analysis.
1982. xi, 92 pages. 28 illus.

Singer/Thorpe: Lecture Notes on
Elementary Topology and Geometry.
1976. viii, 232 pages. 109 illus.

Smith: Linear Algebra.
Second edition.
1984. vii, 362 pages. 20 illus.

Smith: Primer of Modern Analysis
1983. xiii, 442 pages. 45 illus.

Thorpe: Elementary Topics in Differential
Geometry.
1979. xvii, 253 pages. 126 illus.

Troutman: Variational Calculus
with Elementary Convexity.
1983. xiv, 364 pages. 73 illus.

Whyburn/Duda: Dynamic Topology.
1979. xiv, 338 pages. 20 illus.

Wilson: Much Ado About Calculus:
A Modern Treatment with Applications
Prepared for Use with the Computer.
1979. xvii, 788 pages. 145 illus.